In recent years there has been an upsurge of interest in caving at colleges and universities around the world. Large numbers of intelligent, highly motivated, but inexperienced young people are venturing underground. This book is designed for them. It presents the story of caves and the techniques used in their exploration in straightforward but authoritative fashion, drawing on the author's unrivalled expertise in training for cavers.

Nobody becomes a good 'speleo' without understanding how caves are formed, and the initial section describes the processes that have formed labyrinths, deep abysses and chest-constricting fissures in numbers undreamt of by the layman. Half sport, half science, caving is international, so having described what caves are Ben Lyon describes where caves are found, before getting down to the practical business of how to get about in caves, what to wear, and how to become a caver.

Vertical caving calls for special gear and the skills to use it, and a major section of the book is devoted to this most important topic. A cautionary note then spells out the lessons which, well learnt, keep the risks of caving within sensible limits. Finally, the lure of caving soon becomes a desire to achieve every explorer's goal — to tread new ground. How to find, and record, new caves completes this eminently readable book.

About the Author

Malcolm Kensington Lyon (known universally as Ben Lyon) has vast experience and international knowledge of caves and caving. Warden of the Whernside Cave and Fell Centre, Britain's cave training centre, he has for many years been introducing novices to the fascination of the underground world, and the content of the book is based on the training and instruction given on the courses at the Centre.

He is also closely involved in the organisation of caving expeditions; he was joint leader of the 'Mulu 80' expedition which found the world's largest cave, and has caved in many different countries — particularly the USA, Canada, Europe and Malaysia.

Venturing Underground

El Sotano de Las Golondrinas, Mexico.
Nearly 400 metres of free-hanging rope!

illustrated p.3
150m high waterfalls,
in Deer Cave, Sarawak

Dome Pitch, Lost John's Pot,
Northern Pennines, UK

Venturing Underground

the new speleo's guide

Ben Lyon

E P Publishing Limited

Published by EP Publishing Limited, Bradford Road, East Ardsley, Wakefield, West Yorkshire WF3 2JN

ISBN 0 7158 0825 7 (casebound)
0 7158 0826 5 (paperback)

Photography: (unless otherwise acknowledged) Jerry Wooldridge
Illustrations: Barry Davies
Design: Peter Gildersleve

Acknowledgements

The author and publishers are grateful to the following for permission to include copyright material:

Airviews (Manchester) Limited: p. 139
Colin Boothroyd: pp. 11 and 146–147
British Tourist Authority: p. 14
Phil Chapman: pp. 31, 36, 57 and 134
Jeff Clegg: p. 127
Andy Eavis: pp. 2, 6 and 10(2)
Dave Elliot: p. 49
Martyn Farr: pp. 14, 79 and 125
John Forder: pp. 2, 7, 9, 10, 13, 25, 45, 48, 126 and 133
Tim Kelly: pp. 54 and 148
John Gunn: p. 49
Lancashire Evening Post: p. 133(2)
Keith Lewis: pp. 66 and 108
Ben Lyon: pp. 15, 18, 31(2), 154, 158 and 159
Mulu 80 Expedition: pp. 134 and 158
Oxford Scientific Films: p. 18
Art Palmer: pp. 37, 39, 40, 41 and 60
Sefton Photo Library: p. 139
Sheena Stoddard: pp. 12, 51, 53, 62, 71, 72–3, 75, 124, 137 and 154
The Guardian (Denis Thorpe): p. 69
Tony Waltham: p. 16

Printed and bound in Italy by Legatoria Editoriale Giovanni Olivotto, Vicenza

British Library Cataloguing in Publication Data

Lyon, Ben
Venturing underground.
1. Caving – Amateur's manuals
I. Title
796.5′25 GV200.62

ISBN 0–7158–0825–7

Contents

Straw Chamber in Dan-yr-Ogof, South Wales

Revival passage, Mulu, Sarawak. It continues for several kilometr

Smoo Cave, at Durness,
Northern Scotland, is a 'true' limestone cave which ends on the beach

Pitch rigging, Trou Qui Souffle, Vercors, France

Ogof Ffynnon Ddu, South Wales, has the finest cave streamway in Britain

Introduction

From the industrial town in Lancashire where we lived in the 1940s I could see, on the very occasional clear winter's day, just a hint of snow-covered hills. I longed to escape from the smog and mean streets to those far high places. As I grew up, my desire was fulfilled. Weekends and holidays were all spent in the mountains. They lived up to my infant expectations. The views, the winds and clouds — everything about the open air delighted my senses. Nobody could have been less attracted to the idea of exploring caves... not that it ever occurred to me.

When I heard that my kid brother had taken up caving, I was baffled. Why had he done it? Dredging up memories, I remembered the smelly, black air-raid shelters we had played in as children, and thought that caves must be the same sort of disgusting places. All I could remember of a trip into a show-cave at Cheddar was that, afterwards, I was scared by an adder when scrambling around in the gorge. A trip to Gaping Gill, a famous North of England pothole, was arranged while I was working as a mountaineering instructor at an Outward Bound school, but naturally I was not interested — and the tales brought back of mud, soakings and darkness served to reinforce my prejudice!

But then a contradictory note crept into my life. Walking over the Lake District fells I kept encountering dark holes in unlikely places — old mines, long abandoned. Curiosity triumphed over expected revulsion, and before long I was actively seeking out more of these fascinating places. Slate mines, with their spacious caverns, were particularly pleasant to walk through. Evidence of their working — railway lines, clog marks, shot holes — raised ever more questions. Who were the men who had toiled down here, and what else had they dug out of the rock? Answering my questions led to visits to the volcanic graphite mines of Borrowdale, the copper mines above Coniston, and the iron mines of West Cumberland. My forebodings of claustrophobia turned out to be just a fear of the unknown. The joy of contact with rock, learnt from rock climbing, deepened with these ventures into the hearts of the mountains. I was hooked.

Then fate took me to live and work in South Wales, and an introduction to natural caves. These, I learned, normally occur in only one sort of rock — limestone. Surely this meant that one cave would be much like another? Not a bit of it. Although the basic processes involved in cave formation are the same, the finished products come in all sizes, shapes and shades of colour. They are as infinitely varied as any surface scene.

South Wales is renowned for its magnificent caverns. Long, complicated mazes of passages twist and turn beneath the open moorland. Some are dry, their formation completed long ago, monuments to a past landscape. Others have active streams, even rivers, flowing through them. Curiosity, coupled with a sense of adventure sufficient to do something about it, took me swimming along flooded canyons, clambering through boulders, wallowing in mud, and squeezing through narrow gaps like toothpaste through a nozzle. Nobody can deny this physical element in caving; yet to say that caving results in your getting dirty, bruised, soaked and exhausted — though true — does nothing to explain its fascination.

Normal, sane people find caves interesting when they are equipped with concrete paths and electric lights. The speleo is the one in a thousand who just *has* to see what lies beyond. And in return for the discomforts? There for the seeing eye are the beauties of stalactities and stalagmites, crystal pools and natural dams. More subtly, the shapes of the passages themselves, hollowed out in the solid rock, possess a quality no human sculptor could match.

But here a discordant note must be sounded. Caves known to the world at large, and of easy access, often bear sadly little resemblance to their original natural glory. Graffiti mar the walls; any interest the floor once had was trampled out of existence long ago; and broken formations bear witness to both the vandal and the 'collector', the worst vandal of all. Shades of those air-raid shelters.

For fear of spreading this despoliation, cavers tend to be grudging in their acceptance of new recruits to their ranks, and guarded in directing unknowns to their local caves. Thus I and my friends, having little contact with 'real' cavers for a year or two, gradually extended our knowledge, learning from our mistakes. These did *not* include damaging caves — our interest made that unthinkable — but we did run the risk of damaging ourselves. I remember a trip into Ogof Darren Cilau after rain so heavy that the river Usk in the valley below had broken its banks. Sliding head-first into a squeeze, I got stuck, with no way out but back, and then realised that the stream which was trying to get through the same hole as me had belled out my anorak and dammed the passage. As the water rose and started to cascade over my head, I just had time to think how silly a way to die this was before my friends grabbed my legs and uncorked the bottle. We retreated. Such experience is invaluable afterwards — provided one is still alive. A little more initial guidance, such as is given in these pages, should help the reader to avoid predictable perils but still leave ample scope for adventure.

Diccan Pot, Northern Pennines, UK

There are not many potholes* in South Wales, but we read of one in the *Guide to the Caves of Wales* which we had obtained. It was called Pwll Swnd, and was reported to lie in a remote part of the moorland south of the Black Mountains, and to be entered by a sixty-foot shaft. Possessing no tackle for this, but knowing that potholes were descended on flexible ladders, we set to work. The tops of a row of cypress trees in our grounds needed cutting off,

*Confusion often arises as to the difference between caves and potholes. The latter are 'systems' where vertical shafts predominate, or simply single shafts within a bigger complex. Caves are everything else.

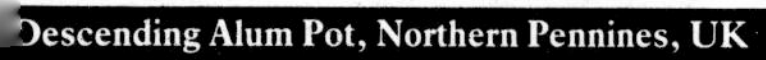

Descending Alum Pot, Northern Pennines, UK

Harwood Hole, New Zealand. Going down down- under

Exit Cave, Tasmania

'he river passage, Krizna Jama, Yugoslavia

A corner of Sarawak Chamber, Sarawak, Malaysia. The biggest known enclosed space in the world

and they made good rungs. Ropes clove-hitched to each side of the rungs finished the job. We set off with our sixty-foot ladder, which weighed about the same number of pounds, and staggered around the mountain all day. Perhaps it was fortunate that we never did find Pwll Swnd!

I learned a little more about potholing before leaving South Wales, but then moved to the Northern Pennines. Here was a land of 'pots'. Black shafts pitted the fells, and, from being an occasional encounter during a day underground, vertical pitches became the norm. Techniques for tackling potholes simply cannot be learned by trial and error, and so I make no apology for the prominence they receive in this book. At the time of my arrival in the north, ladders were the universal method of tackling them. In the ensuing fifteen years a quiet revolution has taken place under the hills: Herculean struggles with enormous heaps of ladders and ropes have been superseded by techniques which, while involving lighter tackle, demand more skill from the individual caver. The transition is not complete, and probably never will be — and nor should it. Cavers have a healthy disregard for arbitrary rules and regulations: any method of progression is acceptable, provided it leaves cave and caver unharmed!

Having become a caver and a potholer, my interest in the caves themselves increased. It was no longer enough simply to 'do' a system. I wanted to know more about the things I could see around me, and what might lie beyond. I joined the Cave Research Group (since merged with the British Speleological Association to become the British Cave Research Group), and was introduced to cave science.

There are those who cave purely for sport, but I am not one of them. Although the techniques of caving are important, so too is an insight into the mysteries of the caves themselves. We are fortunate to live in the golden age of caving. Nobody knows where the deepest cave lies, although one can be certain that it has yet to be explored. Many other mysteries remain to be solved. Just how do living things adapt to a lightless existence? What secrets do sediments in caves contain? How does water circulate through the ground?

Sleets Gill Cave, Yorkshire, U.K.

Yet another aspect of caving is its internationalism. Caves are found in every part of the world, and everywhere attract the same sort of person. Generally rather hairier than the norm, and seldom figures of sartorial elegance, cavers tend to frequent low dives and identify themselves by displaying bat emblems. A certain pallor, unusual in the relatively fit, may also be detected. One aberration I have been unable to explain is that few women are attracted to caving; the only country with a reasonable percentage of female cavers seems to be the USA.

Every four years a grand international event — the International Speleological Congress — brings the world's cavers and cave scientists together. While the latter present learned papers on everything from the possibility of caves on the moon to the age of stalactites, the former swap stories and tee-shirts, and go caving together. In truth there is no clear distinction: the limestone geomorphology expert finds much in common with the simple caver.

Expeditions to distant parts of the globe are an increasingly important part of the caving scene. Their outcome is never known in advance — how can it be, when the actual objective may not even exist? All the reports, written and verbal, from the area to be visited are digested, and the likelihood of further discoveries is weighed up. It may be that a geologically similar mountain near the expedition area is known to contain a big cave. Perhaps there is a bigger catchment area draining into the 'new' mountain. Perhaps the limestone starts at a higher level, and water from it resurges at a lower level, giving the possibility of a depth record. (Many expeditions do indeed claim to be pursuing a depth record — true or not, the claim helps the quest for sponsorship!) Reasons can stretch to the bizarre. An expedition to the jungles of Ecuador went in seach of a cave which, according to first-hand reports, could have been formed only by vaporisation of the rock by a technology more advanced than any known on earth...so spacemen must have done it. The cave was there but, no doubt about it, the smooth outlines were the product of water flow, instantly recognisable to the speleologist!

Gour Fumant — Vercors, France

Caving has been good to me. Meeting farmers on the fells who say, 'You must be daft!', I can reply that at least it makes me a living. Far from separating me from the mountains of my youth, it has taken me to ranges I would never have seen but for the pursuit of elusive holes in the ground. Contact with local people, vital to the caver in foreign parts, has added a new dimension to my life. Scouring alpine pastures for the one shaft in a hundred that 'goes'; prospecting in the Rockies and wondering if grizzlies still lurk in cave entrances; feeling lost in the vast blackness of a cavern in Sarawak — the variety of experience is endless.

However, this is not a book of reminiscences. Instead it tries to give you an idea of what caving is all about, and tell you how to take the first steps to becoming a caver. It will not make a caver out of anyone, but it can, and does, point the way. Practical experience must do the rest. Most people will settle for a vicarious tour of the underworld, through watching the films of Sid Perou and others. Some visit show caves.

Are you one of the few who must experience the real thing?

Fingal's cave, Staffa, Scotland. A typical sea-cave, mostly facade

Stalactites in caves below the Atlantic off the Bahamas are certain evidence of lower sea-levels during the Ice Age

What are caves?

A cave is any naturally occurring hole in the ground. Sometimes those starting with vertical shafts are called potholes, or 'pots' for short. To the layman they are dark and mysterious places, generators of myth and legend. Many of these old stories involve horror and despair — one thinks of the Minotaur's lair, or of Tom Sawyer lost in a later labyrinth of Mark Twain's mind — so, in the public consciousness, caves have become places to avoid at all costs.

Caves *can* become horrible places when interfered with by human beings. Holes in the ground have been used to dispose of bodies, including the wholesale dumping of corpses down open potholes, since time immemorial. They are perhaps the last aspect of the natural world to be regarded as wild and awful places. This reputation worries cavers not one bit. It keeps the population at large from discovering that in reality caves are places of fascination and strange beauty!

'What does the abyss hold?' Hunt Pot, Northern Pennines, UK

Ice formations inside the lower entrance of Lamprechtstofen, Austria

Cave formation

Caves form in various ways. Perhaps most familiar are sea caves, those dark clefts formed where the pounding waves have picked out the weaker lines in hard-rock cliffs. Fingal's Cave, in the basalt columns on Staffa, and the numerous caves along Cornwall's granite coast, are typical. But despite old smuggling tales, sea caves rarely amount to much, usually growing smaller from the entrance in, closing up completely beyond the waves' reach. Occasionally, though, the tremendous pressure created when a wave is compressed into the narrowing end of a cave is sufficient to blast a hole through to the clifftop, creating a 'through-cave'.

In a few locations, sea and undersea caves are long and complex, as for instance under, and in the sea around, Bermuda.

These are not true sea caves, though. They were formed long ago by the solution processes described below at a time when sea-level was more than a hundred metres lower than it is now, because so much water was locked up in the glaciers of the ice age. They form a special type of 'fossil' cave.

Less well known are lava caves. Jules Verne 'invented' these for his *Journey to the Centre of the Earth*, but outside fiction they are only found close to the surface. In certain types of lava flow, the surface layer cools to form a solid roof over the still molten rock beneath. This molten rock flows on leaving passages which can be kilometres long, adorned with glass 'stalactites'. Only recently has the extent of such caves been realised. Many are found on volcanic islands, the best known being in Iceland, Hawaii and the Canary Islands.

Lava Cave, Hawaii. The tube left after the molten lava flowed on

Many lava-tube caves are of recent origin, but the youngest and fastest-changing of all caves are those formed within the ice of glaciers. In summer, glaciers have meltwater streams on their surfaces which are swallowed by mysterious shafts, the water emerging in the form of rock-laden torrents from the glacier snouts. In between lie constantly changing networks of passages, little explored, and only recently starting to attract the serious interest of cavers.

Sea, lava and glacier caves are, however, a mere *hors d'oeuvre* to the dominant type of caves — those formed by the dissolving away of rock. The main type of soluble rock is limestone, which covers about one twelfth of the world's land surface. Caves formed in it have an intimate relationship with the structure of limestone, so it is worth looking at the way that the rock is created.

It usually consists of the skeletal remains of countless sea creatures — some large, but many very small — which collected in layers beneath warm seas, often in lagoons bounded by reefs. The process continues today off the coast of Florida, in the Indian Ocean, and in many parts of Australasia. Apart from coral-reef limestones, most limestones have, therefore, been laid down horizontally on the sea bed.

From time to time the climate, or the sea level, has changed, resulting in a break in the deposition of the limestone, and splitting the resulting rock formation into definite layers, or beds. Sometimes there is a mere crack between beds; sometimes a layer of insoluble mud which has since been compressed to become shale. These junctions between beds of rock are known as bedding-planes. Another sort of crack results from the stresses and strains of geological time: called joints, such cracks occur in a criss-cross pattern, usually at right-angles to the bedding-planes. So regularly do these types of crack occur in limestone that geologists use a brickwork pattern on maps to indicate it.

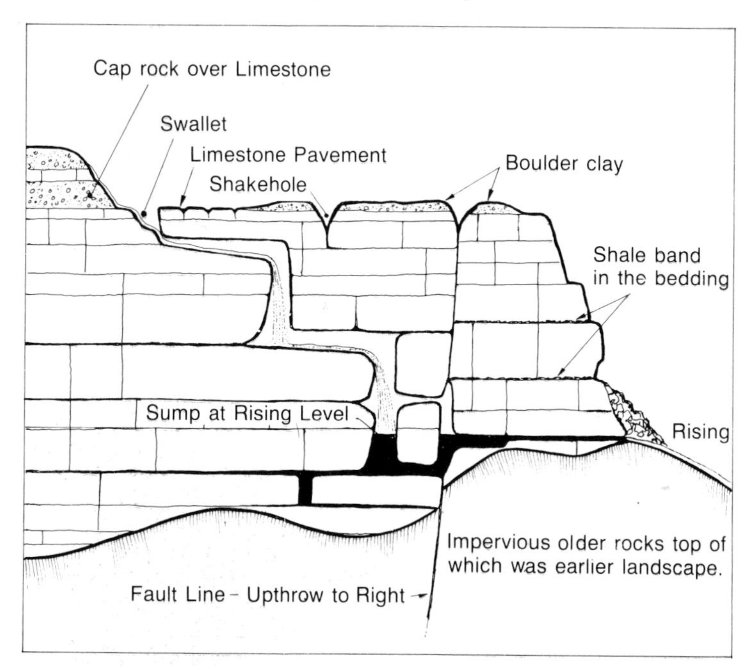

Section through limestone. Diagrammatic representation of the Great Scar limestone of the Three Peaks area, Yorkshire, UK

Sometimes limestone is uplifted from the sea bed with very little alteration. The Mississippian limestones of Kentucky have titled by only a fraction of one degree, and those of the north of England tilt (or, to use the technical term, dip) by only two or three degrees. In Somerset, however, the Mendips are formed by

Limestone — the stuff from which caves are formed — is still being deposited in tropical seas

illustrated p. 19
Progression may involve 'bridging' above a deep trench

Typically sharp-edged solution channels in sloping mountain karst

limestone beds arched up (folded) so that they dip at around 45 degrees. Even this is as nothing compared to the folding and faulting (shearing of the rock) which occurs in the big mountain ranges. Recently I was caving in limestones in the Rockies which had been folded upside down, so that the oldest rocks now overlie newer beds! The inclination of the limestone affects the type of caves found in it, as we shall see later.

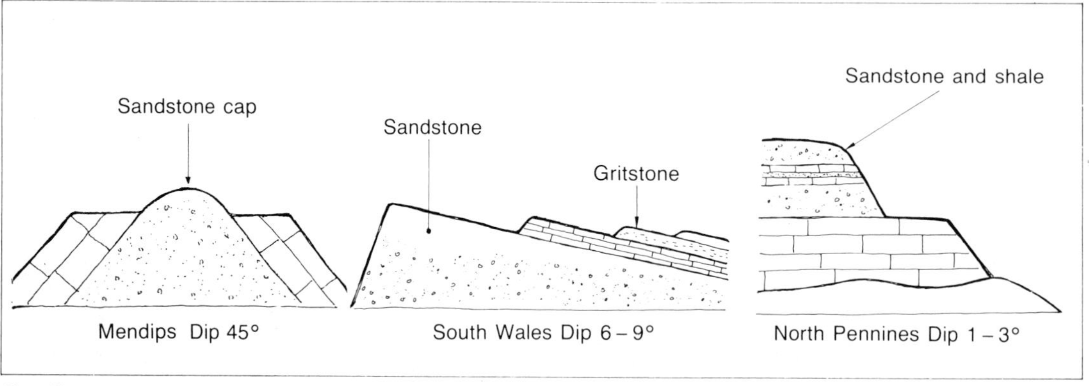

Simplified geology of different caving areas

Seen under the microscope, limestone is a mass of interlocking crystals. This tight structure means that it is imporous, so water can move only through actual cracks in the rock. Its chemical composition is almost pure calcium carbonate, which is slightly soluble in rainwater. However, because the rock is imporous, solution can occur only where water seeps through the weaknesses — the joints and bedding-planes. When rain falls on a bare limestone surface it runs into the joints, opening them up, so that deep cracks form around pedestals of solid rock, giving a pavement surface. Below these the water running down the joints encounters bedding-planes, and is channelled along them. Gradually more and more water is concentrated into fewer and fewer pathways, until it has sufficient volume to dissolve passages big enough for people to enter.

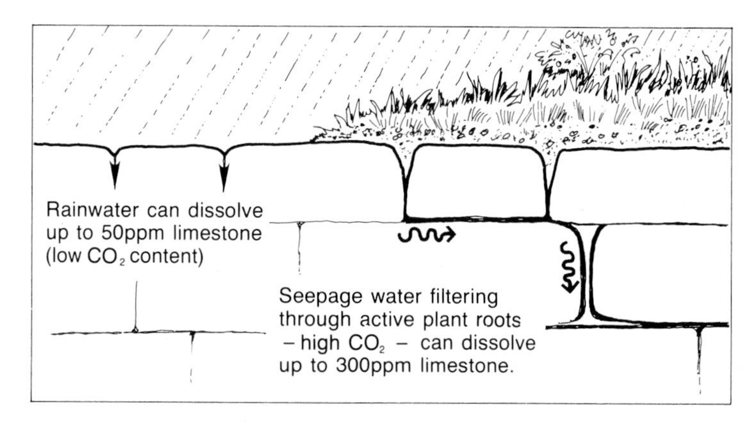

The solubility of cave water

Stream water draining off acid soils may dissolve 50 – 60 parts per million of limestone (calcium carbonate)
Rain water falling directly on to bare limestone will have a low CO_2 content and dissolve less than 50 p.p.m.
Seepage water filtering through active plant roots has a high CO_2 content and can dissolve up to 300 p.p.m. of limestone

The actual solution process is very complex, and the complete chemistry well beyond the scope of this book. However, a simplified version will suffice for basic understanding. Rainwater contains carbon dioxide, absorbed from the atmosphere. This turns rainwater into a weak acid, capable of dissolving up to about fifty parts per million of calcium carbonate. In addition, the soil around active plant roots contains much more carbon dioxide than the atmosphere, so water seeping down through it becomes more strongly acidic, and capable of dissolving up to about three hundred parts per million of calcium carbonate. The more carbon dioxide in solution, the more aggressive the water/acid, and so the more limestone can be dissolved.

Flat-bedded limestone pavement, showing solution of joints. Ingleborough, Northern Pennines, UK

Types of Cave

In the initial stage of cave development, all the cracks in the limestone are full of water, which is moving slowly through, fed by rain or 'sinking' streams and emerging at springs; passages formed in these conditions are known as 'phreatic'. Because more water flows through wider cracks than narrow ones, it is the wider ones that are corroded, or dissolved, the most. This makes them able to take yet more flow and, from an initial network of multitudes of tiny solution channels along joints and bedding-planes, a few grow large and become real cave passages. In these phreatic conditions the water can erode the entire outline of the passage, which tends to be smooth and rounded, although still reflecting the nature of the weakness which started it.

The slow-flow conditions that obtain when passages are completely full of water lead to very large current-markings, or scallops, on the walls. These indentations, sometimes as large as soup plates, have a sharp lip at their upstream end. Thus, even if the water that formed a passage disappeared long ago, it is still possible to chart its progress. This directionality means that in small passages it is easier to crawl 'with' the scalloping than against it — a good example being Hensler's Crawl in Gaping Gill.

illustrated p.22
Gour pools of large size. Gouffre Berger, France

Phreatic features come in many sizes. Peak Cavern, Derbyshire, UK

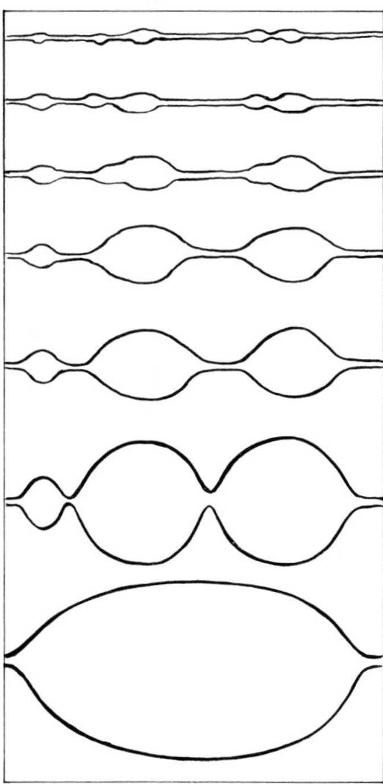

Phreatic passage development

Water seeping through the bedding in a direction perpendicular to the diagram dissolves more rock where the crack is widest. This draws more flow to the tubes so created until eventually one large tube is made

A small phreatic tube — formed in water-filled conditions

Soup-plate sized 'scallops' indicating former slow phreatic flow
Link Pot, Northern Pennines, UK

The depth of phreatic systems can be hundreds of metres, with water penetrating far below the water-table before working its way back up to resurgence level. (In the flooded passages, water can flow upwards as well as down.) During their formation, such caves are accessible only to cave divers, and even then only partially so. Although dives of several thousand metres in *length* have been made, the deepest so far explored is to 160m in the Fontaine de Vaucluse; it continues downwards unexplored. However, the landscape undergoes constant change, and vast numbers of caves formed under water have now been drained and are accessible to the caver. Sometimes this has been due to the raising of land-surface levels and the downcutting of rivers running through the area. The recent (geologically speaking) ice ages have dramatically aided the de-watering of phreatic caves. Glacial scouring has deepened valleys in mountain and hill areas throughout much of the world, leaving old spring levels high on the mountainsides.

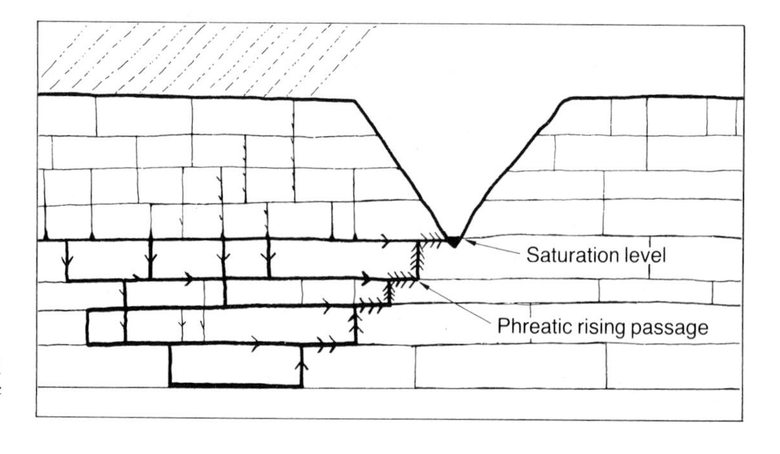

Deep phreatic flow

Water falling on limestone seeps down any available cracks until it reaches saturation level. Below this level the water will still move downwards under the pressure of added supplies from above, but will then move towards the nearest available resurgence point. This feeds an upward-flowing tube. Subsequent lowering of the resurgence level, due, for instance, to glacial down-cutting, may drain off the old phreatic tubes

Some of the old water-filled passages are left just as they were formed, their rounded outlines explaining their origin as clearly as a museum label. Such passages, if left high and dry, are called fossil passages. Others still have water running through them, this time acting directly under the force of gravity to reach the lowest level as efficiently as possible. These conditions, with water flowing in the base of the passage and air-space above, are known as 'vadose'. Unlike the general ballooning-out of passages created in phreatic conditions, vadose conditions lead to downward cutting in the floor of the passage, as this is the only part with which the water is in contact. Vadose trenching is a common feature of caves all over the world. Free-flowing streams tend to move faster than those occupying the whole passage section. This gives rise to much smaller scalloping and also allows corrasion to play a larger part in the development: the passage floor is worn down by the bashing it receives from solid material — from sand grains to boulders — carried by streams.

Vadose trenches may be very narrow and winding, or quite enormous. The smallest are too tight to enter; the largest are huge underground canyons. The biggest I have been in, in Yugoslavia, is over 70m high and carries a major river in its depths. Large or small, vadose caves have a similar cross-section. Up in the roof

Downcutting in vadose conditions produces typical trenched passages. Lancaster Hole 'main drain', Northern Pennines, UK
NB Small scallops

Progression may involve 'bridging' above a deep trench

the original phreatic tunnel is left as a 'half-tube', and below this the vadose slot cuts down. Such combination passages are known as keyhole passages, for obvious reasons. It is often necessary to move through them at the top level, trying to avoid a slip into the slot beneath. Bridging techniques (see p.74) are invaluable here. Conversely, moving along a narrow vadose trench calls for the suppleness of a contortionist. The moving water swirls around each bend, eroding the outside wall each time to make more and more exaggerated looping as it cuts down. Each twist and turn calls for corresponding movements from the caver.

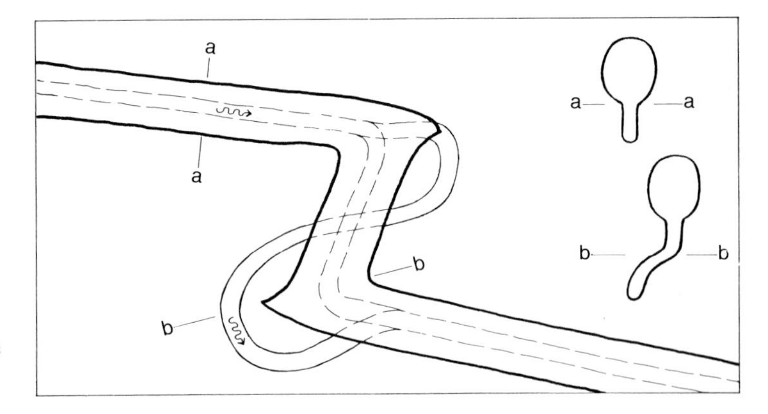

Once the original phreatic tubes are drained, subsequent development is concentrated in the floor of the passage — giving rise to meandering 'vadose' trenches and a typical 'keyhole' section

illustrated p.27
Breakdown and blockfall reveal the beddings and joints in limestone. Pant Mawr, South Wales

Cave morphology

Just as buildings can exhibit astounding variety despite being made from the same materials, the same basic processes produce caves of quite different shapes. The caves of the UK illustrate this very well.

In South Devon, on the edge of Dartmoor, there are caves formed in submerged conditions in very old and contorted limestones that are like the inside of a Gruyère cheese — rounded caverns interconnected by rounded tube passages. Apparently random, these holes faithfully show the pattern of the cracks in the rock which water could exploit. No streams have flowed in most of these caves since the local river-valley level dropped enough to drain them.

In the Mendips, in Somerset, there is a sandstone capping which collects streams and delivers them onto limestones dipping at around 45 degrees. The stream caves drop steeply down the bedding, sometimes eroding trenches, elsewhere a series of water slides and pools. As all the weaknesses — both bedding-planes and joints — are inclined, there are few vertical shafts. The streams quickly reach saturation level, around 150m below the surface, and thereafter must follow a switchback route to their resurgences, up phreatic tubes along a joint line, and back down the next bedding-plane. The crest of each zigzag is trenched into by free-flowing water, allowing the caver access, but each dip is still submerged — hence the long sequences of sumps for which the area is famous.

The streamway, Ogof Ffynnon Ddu, South Wales

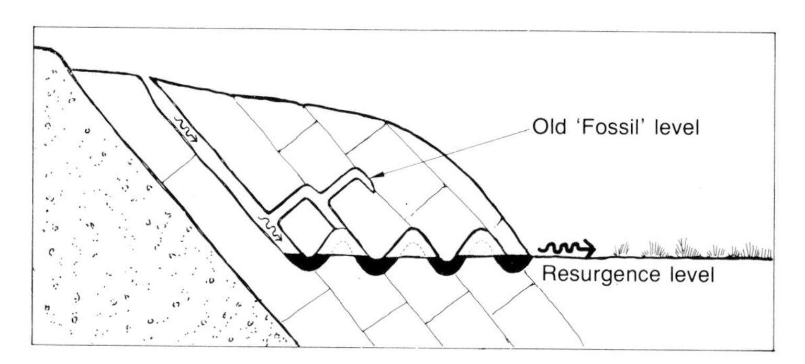

In the steeply dipping limestones of Mendip water quickly reaches resurgence level. It then proceeds on an up-joint/down-bedding switchback which gives the caver a series of sumps to negotiate

Limestones dipping at just a few degrees — like those found to the north of the South Wales coalfield — allow water to find its way downwards along gentle inclines, with the minimum of vertical drops. These are the right conditions for long, uniform vadose trenches to cut their way down as streams flow along down-dip passages. The best example of all is the streamway of Ogof Ffynnon Ddu, some thirty or more metres high and over four kilometres long.

When the beds of limestone lie horizontally, or nearly so, as in the northern Pennines, cave passages formed in bedding-planes make very poor routes for water to move to lower levels. Hence broad, shallow passages are formed by the slow-moving water until it can exploit a joint and fall vertically to a lower level, with another horizontal passage leading to another shaft. The initiation of vadose trenches here begins at the lip of pitches, the water

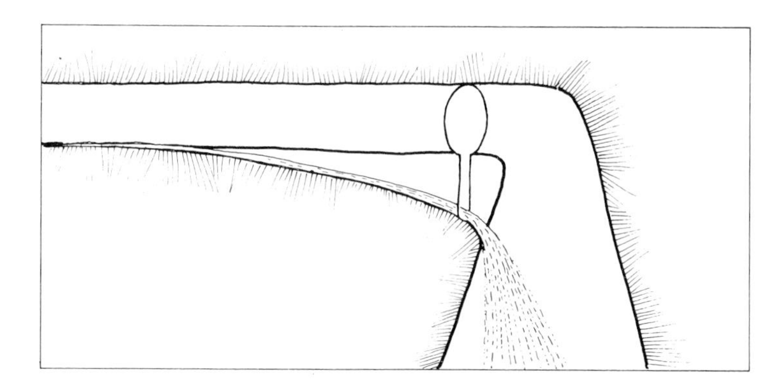

Water flowing over the lip of a shaft accelerates and cuts a deep, narrow slot in the passage floor

accelerating over them to give particularly deep and narrow slots. A knowledge of this process is of use to the practical caver. When a stream passage suddenly starts to cut down in its floor, a pitch is rarely far ahead. A scramble up to the old, phreatic level in the roof enables the caver to traverse along above the tight streamway, and rig the pitch above and beyond the falling water.

Both phreatic and vadose development leave nicely sculpted and rounded rock surfaces. However, once water has opened up the rock weaknesses, the collapse or breakdown of passages along those lines of weakness may follow. The cave no longer has a smooth outline, but becomes sharp and squared-off, with boulders littering the floor (see page 27). Such collapse can continue until the passage is entirely blocked or the surface is reached.

The water which enlarges caves can also carry stones and silt to block them. Kingsdale Master Cave, Northern Pennines, UK

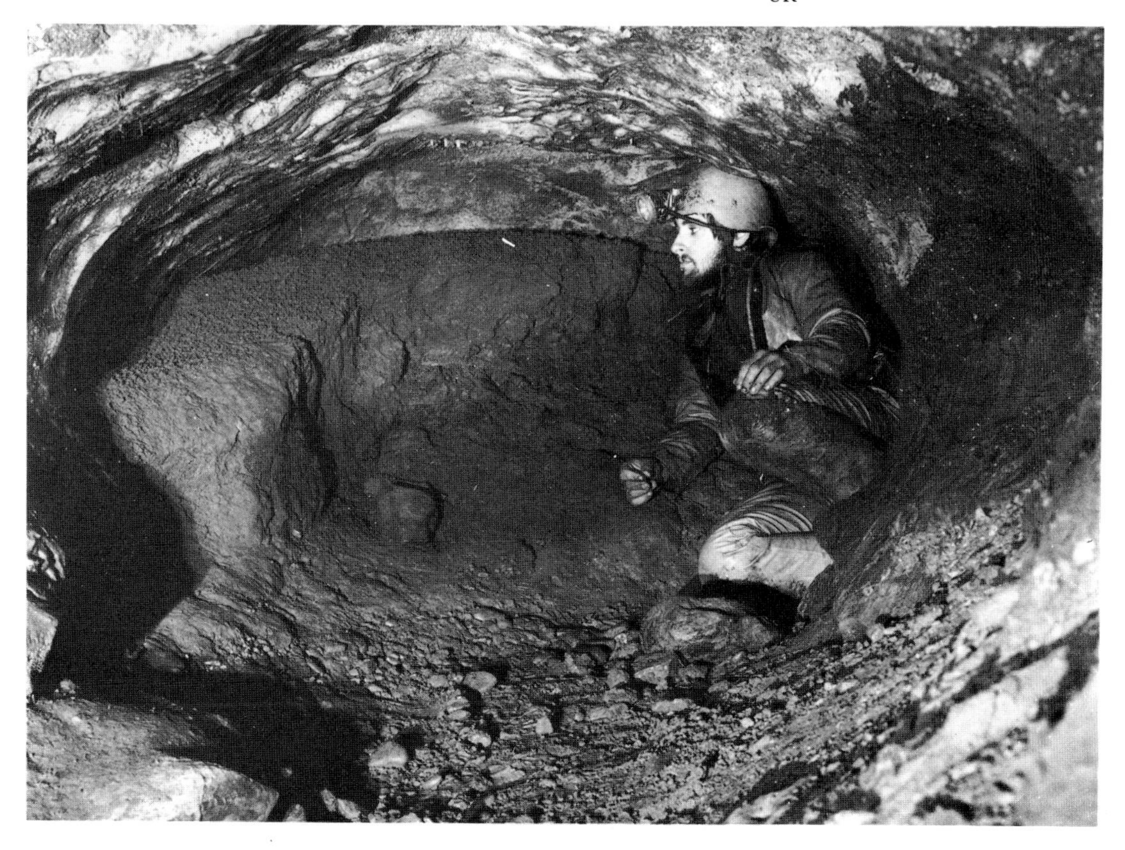

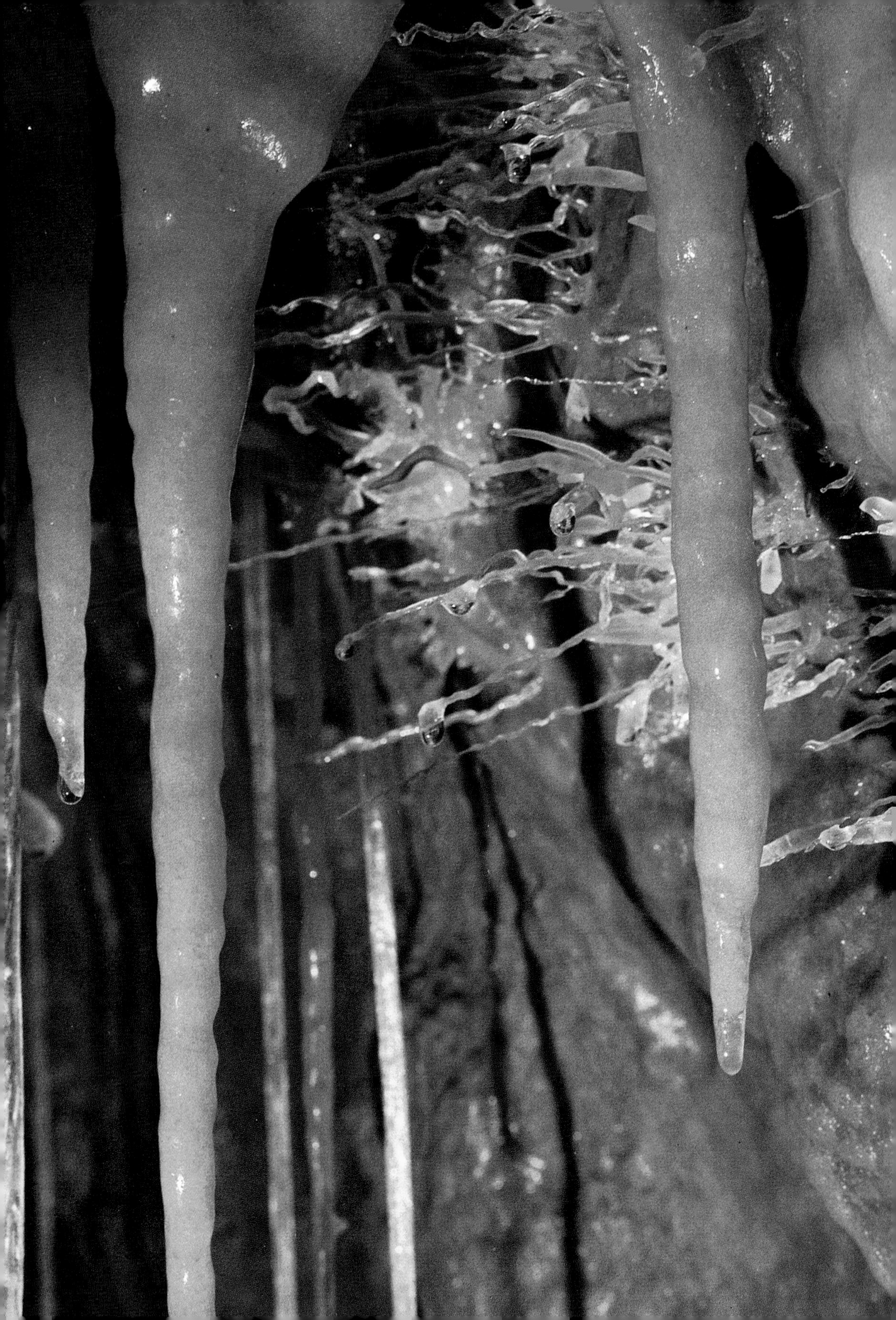

illustrated p.30
Helictites, Ogof Ffynnon Ddu, South Wales

Bats should be disturbed as little as possible

Living its entire life-cycle in blackness, this troglobite crab could not exist outside its cave

Water flowing through a cave may bring with it silt, pebbles and boulders and leave them there. During the ice ages, when vast quantities of solid material were being redistributed over the landscape, many caves became more or less completely filled with such infill. Some passages are still blocked by it; in others it has been washed out again.

Stalactites and stalagmites

The last stage in the development of a cave gives rise to what most visitors to show-caves go to see. When limestone is precipitated from solution it can form beautiful and amazing formations. The deposition occurs when carbon dioxide in seepage water 'gasses off' and calcite crystals are precipitated.

The commonest stalactites are 'straws'. Though common, they can be among the most beautiful. When water seeps from a crack, a thin ring of crystal is deposited around the circumference of the drop where it clings to the roof. The drop falls, another takes its place, and the thin ring gradually becomes a hollow tube. 'Straws' can grow to a length of three metres if undisturbed. They are unique among cave forms in that they always have the same diameter — that of a water drop.

Tapering, or 'carrot', stalactites often start as straws, filling out when the central hole becomes blocked and water seeps down the sides of the straw. Nearly all stalactites draw to a water-drop point at their lower end — unlike the stalagmites which grow up towards them. The latter, formed as drops splash down and flow outwards, present a more stumpy appearance. They can grow to enormous proportions, the only limitation on their height being a junction with a stalactite or the passage roof to form a column.

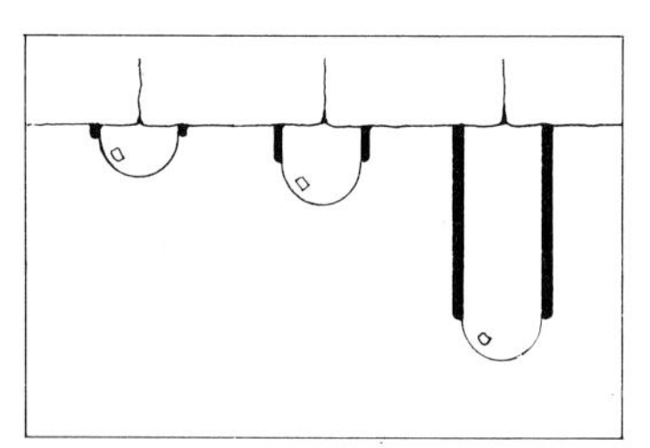

'Straw' formation. Seepage water deposits a 'rim' of calcite around the edge of the drop. This grows to give a hollow rock tube

A chamber roofed with 'straws'. White Scar Cave, Northern Pennines, UK

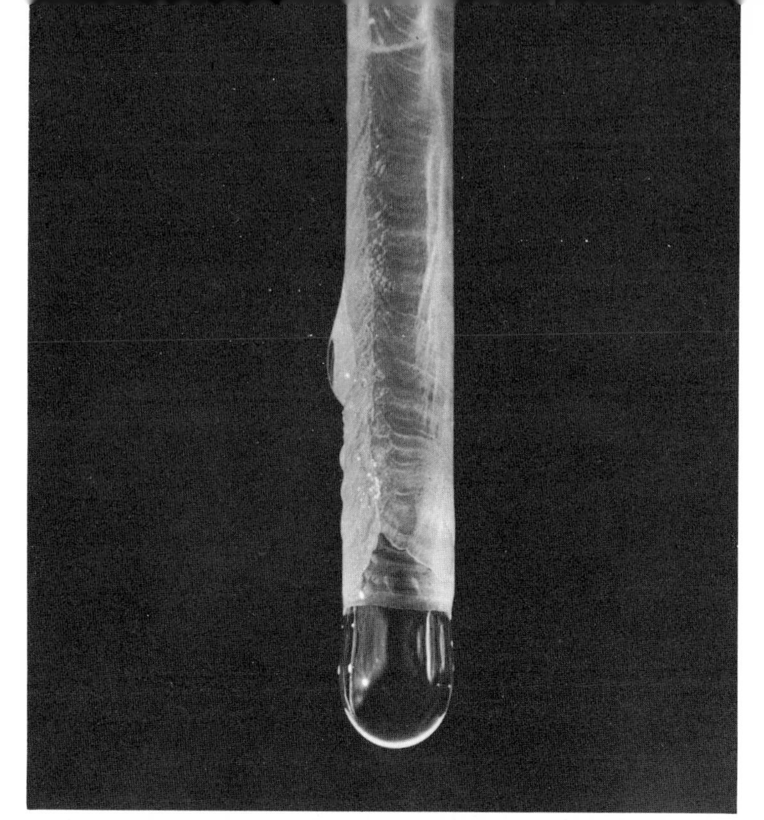

Straw stalactites have the diameter of a water-drop

Stalactites and 'mites. They will eventually join to form columns

More common than free-hanging stalactites and their complementary 'mites is flowstone — calcite which simply coats the walls and floors of passages. It often creates a chinese-terrace effect, with pools dammed behind barriers of calcite. These gours or dams can be formed in streamways as well as on slopes. One reason for their formation is that more excess carbon dioxide can be given off when water is curved shallowly over a lip than from the flat surface of a deep pool. Hence more deposition occurs on protrusions, which gradually become horizontal dams. They can be tiny — a mere fraction of a centimetre — or big enough to create pools deep enough to drown in.

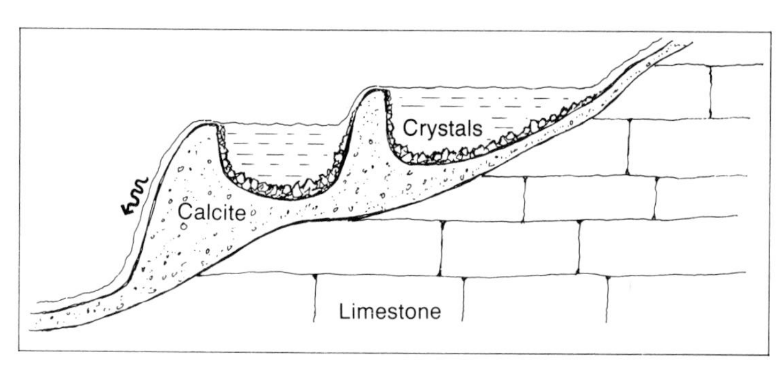

Gour pools. Water gives off CO_2 and must deposit some calcite. More CO_2 is given off in the thin film over the lip, therefore more deposition takes place here — hence terracing develops

It is relatively easy to explain the formation of the kinds of crystal deposits we have so far mentioned. Another type, the helictite, is more mysterious. Helictites vary from single twisted rods to complex intertwined masses, but all appear to defy gravity, growing in any and every direction. One theory has it that they form where water seeps in minute quantities, but under high pressure, from a tiny crack; calcite is deposited, and as the helictite grows a capilliary tube continues to allow the high-pressure seepage to flow to its tip, which continues to grow. Well, maybe. My own favourite theory is that tiny ionised water droplets produced in the spray of waterfalls are attracted to oppositely charged points on the cave wall.

Life in caves

All the formations we have discussed are made of hard, crystalline calcite. However, in many caves white deposits of moonmilk (looking very much as their alternative name, 'cave cauliflower', would suggest) can be seen adorning the roof. Moonmilk has a floury texture; it is limestone 'fixed' by bacterial action. Most formations are white, or nearly so, but they can be given a wide spectrum of colours by minerals. That said, a quite common translucent orange colouring which used to be thought to be due to the presence of iron salts has now been shown to be also the work of bacteria!

Many bacteria, because they need no light, can live in caves. So can many higher animals, provided they have a source of food. This may be brought in by streams or by animals which spend only part of their time in caves. Best known of the trogloxenes are bats, which use caves for roosting and hibernating but sally forth at dusk to feed. They bring back a rich fund of organic material in the form of their faeces (guano). This allows a complete world of

Curtain, Withyhill Cave, Mendips, UK

smaller creatures to live out their life-cycles in complete darkness — the true troglobites. They have adapted from surface-dwelling species to such an extent that many are now quite different: eyes have disappeared, vibration sensors have developed, metabolism has slowed down.

In temperate and cold countries the numbers and varieties of cave life are limited. In the UK, for instance, the largest troglobites are small shrimp-like creatures called *Niphargus*. The much larger blind white trout seen in cave streams are simply river-dwelling fish which have been washed into the cave. Lack of light has caused pigment to disappear and eyes to become useless, but no positive adaptations which would enable the fish to live their entire life-cycle underground have yet been made.

Cave cricket, numbered by a cave biologist during a study of the creature's habits

The best place to see such adaptations is in warm and tropical caves. Here are found eyeless fish, newt-like creatures called salamanders, and numerous types of insects and spiders. The underground world has provided a stable environment in which the modern cave-adapted species are sometimes the only living ancestors of creatures long since extinct above ground. This makes them extremely interesting to science; the importance of the preservation of caves as repositories of the past becomes obvious.

Dating

The age of stalactites can now be measured. As a stalactite forms, two isotopes of uranium are present in the calcite in a fixed ratio. This proportion alters with time, as a result of radioactive decay, and thereby an inbuilt 'clock' is provided. This has proved a powerful tool for the archaeologist (studying remains of cave dwellers embedded in calcite) and for the geomorphologist, who can correlate dates discovered in stalactite tests with, for instance, specific phases in the ice age. On a caver's level, it means that a cave must have been formed before the stalactites in it — and, in the UK at least, this has already led to many surprises. Tentative ages had been given for many caves based on estimates of the amount of limestone being removed from them in solution by water passing through, and it appeared that some at least had begun to form after the most recent ice age. But, since stalactites in some of them are now known to be much *older* than the ice age, the cave must also predate it!

illustrated p.37

Upper-level passage in the Mammoth Cave System, Kentucky; typical of many of the large main passages in the system, it is a high, wide canyon partly filled with sediment and breakdown. This is Dyer Avenue in the Crystal Cave section, which was once under private ownership and open for tourists as late as 1961. Remnants of trail improvements are still visible

I must add, though, that stalactites do not always take a long time to form. Their growth depends on the prevailing water and air conditions, and on the amount of calcite and carbon dioxide in solution. Straws *can* grow centimetres in a year, in exceptional circumstances, but on the other hand it may take thousands of years.

Nothing is simple in caves. That is one of their charms!

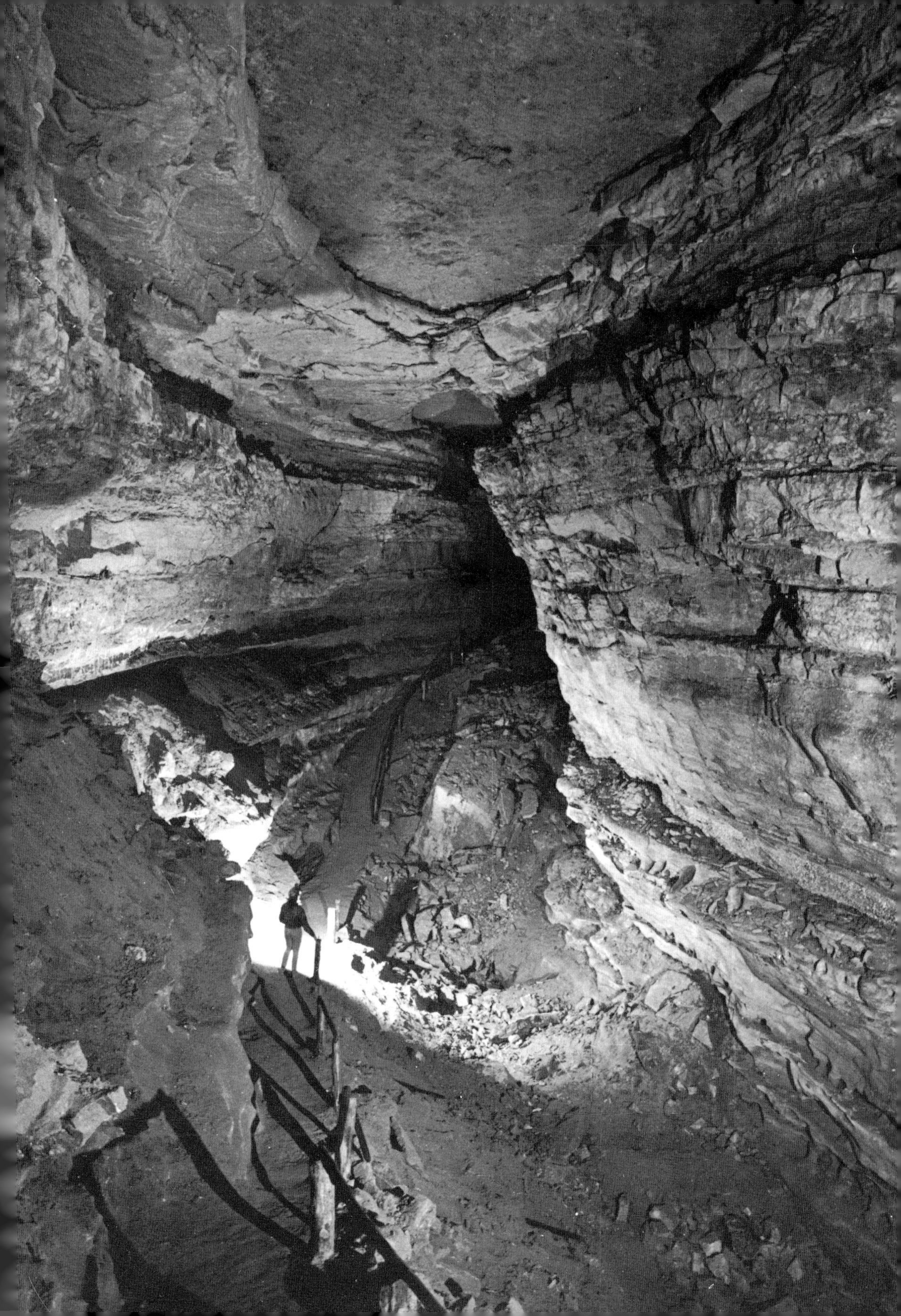

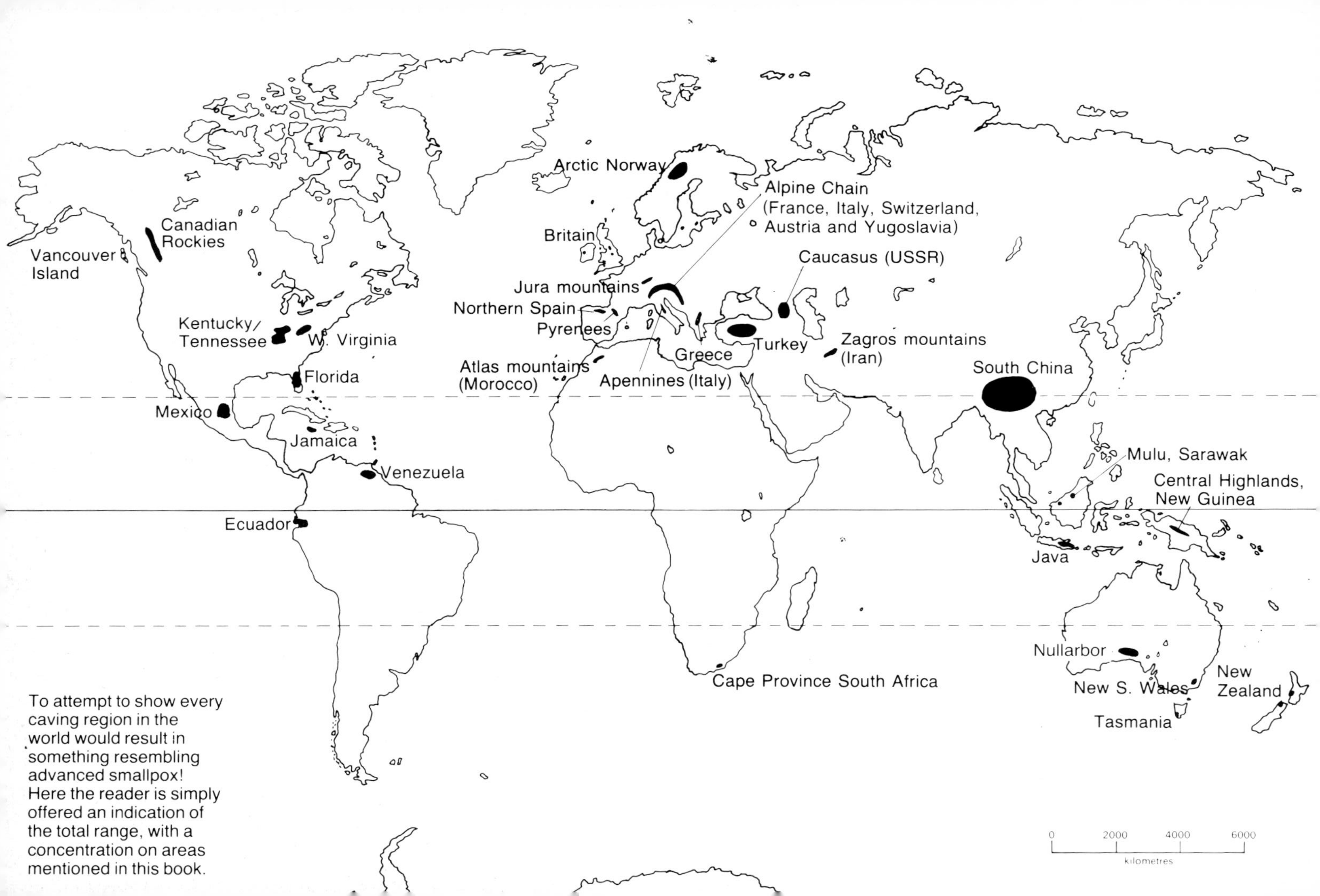

To attempt to show every caving region in the world would result in something resembling advanced smallpox! Here the reader is simply offered an indication of the total range, with a concentration on areas mentioned in this book.

Where caves are found

The world scene

Cavers have found and explored only a fraction of the world's caves and only in the last dozen or so years has the prospecting of some of the great limestone ranges in remote parts of the globe even started. Each year expeditions push back the frontiers; and the score sheet of the largest and longest caves changes so rapidly that no publication can keep up to date. However, the longest known cave-system is unlikely to yield its crown for many a year, if ever!

The record-holders

In Kentucky, USA, the Green River meanders through hilly country, snaking round the bases of interlocking ridges. Only rarely does the wooded ground reveal a glimpse of the underlying limestone, but this gentle landscape covers the most amazing network of caves. The passages lie at six different levels in the almost flat-bedded rock; they vary in size from airy canyons to rat-holes too small for most to venture into. Only at the lowest level does one meet water nowadays. As the Green River cut downward, so the active cave level matched it, leaving hundreds

The larger passages of Mammoth Cave are interconnected by nasty, low crawlways and maddeningly confusing canyon complexes. The temperature is about 14°C, warm enough so that wetsuits or oversuits are not usually worn, except sometimes where long bouts of total immersion are expected. This passage is a tributary of Mystic River in Mammoth Cave, a section that was discovered in 1973.
(Passage was virgin when picture was taken.)

of miles of 'fossil' cave, occasionally intersected by complex shafts where present-day rainfall drains off the sandstone cap-rock. Parts of the caves have been known for centuries, with the Mammoth Cave being the first major system explored for any length. Then knowledge of the Flint-Ridge system, some miles away, started to 'grow', initially as a result of the solo efforts of Floyd Collins, seeking for a show-cave to eclipse Mammoth. His slow death in 1925, trapped in a Flint-Ridge cave, led to national fame — alas, posthumous. His body, suitably coffined, became the chief attraction of the Flint-Ridge caves, and rests there to this day. Eventually, in 1974, a crawlway connected Mammoth and Flint-Ridge, and the combined network of caves has been surveyed to over three hundred kilometres — and is still growing! In this cave the only limitations on exploration are bruised knees and exhaustion.

One of the many tubular trunk passages in the Mammoth Cave System — Grand Avenue in the Colossal Cave section —which extends for several kilometres without interruption

The USA boasts many other long and complex caves, the most sporting perhaps being in West Virginia, with potholes ('pits' is the local term) as well as passages. But for the largest chamber in the USA we have to travel to the Big Room of Carlsbad Caverns in New Mexico. This has a volume of around four million cubic metres, and was, until 1981, a contender for the title of world's largest cavern together with La Verna in the Pierre St Martin system in the Pyrenees.

Awesome though these are, they have been dwarfed by the finding of Sarawak Chamber inside Mount Api in Borneo. The first action of cave explorers is (or should be) to survey their find. The team of British and Malaysian cavers who found Sarawak Chamber spent more than twelve hours just surveying in total blackness, apart from occasional glimpses of soaring wall on one side, before they realised that they were actually creeping round the edge of an enormous empty space. I had the great good fortune to be in the party that followed to complete the survey. It showed a volume of at least twelve million cubic metres! Just climbing from one side of the chamber to the other involved a vertical ascent of 300m or so, on a boulder-and-scree slope never consolidated by wind, rain or vegetation. Perhaps nobody will ever really see this hollow mountain — certainly our lights, even the thousand-watt filming lights, just created an island in the darkness.

illustrated p.41
Main passage of Ogle Cave, Carlsbad Caverns National Park, New Mexico. This and other caves in the Guadalupe Mountains, was apparently formed by sulphuric acid from oxidation of rising H_2S from oil-field brines. The stalactites and stalagmites were formed in the usual way

There is no doubt that many of the most exciting finds of the future will come from the tropics and semi-tropics. Cavers from Austin, Texas, travelled south to Mexico to descend the greatest known 'pits'. Deepest of all is El Sotano, at 410m. In the same area — the Sierra Madre — is Las Golondrinas, which bells out from its narrow top to give a floor area of nearly three hectares some 350m below. It was in such 'pits' as these that US rope techniques were developed, with emphasis on long, uninterrupted descents and rapid rope-walking methods of ascent. Very important when you have to climb a rope hanging in free space for nearly a quarter of a mile!

The deepest cave systems in the Americas are in Mexico, and cavers from the USA, France, Belgium and Britain have all made big discoveries in recent years. The pattern of weekend caving in one's home country, plus a yearly expedition to foreign parts, is becoming a regular feature of the cavers' world. Mexico is attracting much of this expedition attention, and, on the other side of the world, South-east Asia and New Guinea are becoming new caving 'Meccas'. Western cavers eagerly await the day that access is allowed to the huge karst areas of Southern China.

Traverse line — Gouffre Berger, Vercors, France

British cavers have opened up more than a hundred kilometres of huge caves in the Mulu National Park in Sarawak — Sarawak Chamber, already mentioned, is, although the largest, just one of many giant caves there. The quality of these caves is such that they have become the yardstick against which other claims to big caves have to be measured. To explore them involves none of the usual crawling, or special clothing to keep the cold at bay — just boots, shorts, and a light. The hazards of these and other tropical caves are flash-flooding, heat exhaustion and possible attack by strange beasties, including snakes!

Not all the hottest caves are in the wet equatorial region. A seemingly unlikely setting for long caves with extensive flooded sections lies beneath Australia's Nullarbor Plain. A friend of mine nearly passed out from the heat when climbing a rope out of one of these.

The European continent

Back in Europe, the chain of mountains from Greece, through Yugoslavia, Austria, Italy, Switzerland and France, centred on the Alps, contains more known deep systems than anywhere else. Indeed, if the Caucasus (to the east) and the Pyrenees and mountains of northern Spain (to the west) are added, the zone contains nine-tenths of the world's deepest caves, including the current 'title-holder', the Gouffre Jean Bernard in the Haute-Savoie of France, at –1455 metres.

Greece and Yugoslavia

Greece has shafts to rival those of Mexico, like the Epos Chasm and Provatina, both 'bottomed' first by British expeditions. Yugoslavia contains a higher percentage of speleogenic limestone than any other country, and in the northern province of Slovenia cave research has been a respectable science for more than a century. It is here that the classic features of cave-bearing scenery were first decribed, and the general name 'karst', to describe waterless limestone landscape, comes from the Carso, a area in the hinterland of Trieste.

The Postojna system is world-famous, its show-cave section visited by underground railway! In the river passages below are eyeless cave salamanders (the only ones from the Old World, though there are several species found in the Americas). Slovenia has, too, given the name 'doline' to large, closed limestone depressions. Another local feature, poljes, are uncommon elsewhere. They are valleys, floored with soil and totally ringed by limestone hills, which become lakes in winter and drain slowly in spring to allow crops to be grown before the next winter's inundation.

Austria

The Julian Alps in the far north of Yugoslavia overlook Austria, which has many fine cave areas. Two famous ones that I have visited are the Tennengebirge, south of Salzburg, and the Leoganger Steinberge, not far from Hitler's Bavarian hideout at Berchtesgaden.

The Tennengebirge contains a number of fine caves. Perhaps the most remarkable is the Eisriesenhöhle. Entering through a relatively small fossil opening halfway up the mountainside, the visitor is confronted by a descending glacier! It has been estimated that the first kilometre of this cave contains thirty thousand cubic metres of ice, a product of the freezing wind which is sucked into the entrance during the winter. After this first kilometre the air has warmed up, and the remaining 39km is ice-free.

Climbing up through the Lamprechtstofen involves numerous iron ladders and rope traverses — the ultimate in fixed aids!

The Lamprechtstofen, in the Leoganger Steinberge, is likewise entered by a lower entrance, this time at valley level, not far from the village of Weissbach. It is another world-record holder, this time for the cave involving the greatest climb. It has been ascended to a height greater than 1000m above the entrance. Like many Alpine caves it is best tackled in winter, when the surface is snow-covered and sudden flooding less likely. Getting to the top end involves traversing along fixed wires, rafting along canals, endless climbs up not-so-fixed ladders (the Austrians are very keen on fixed aids), bivouacs, and the passing of horribly loose boulder chokes. These last tend to occur in *klamms* (the Austrian word for faults). They are some of the most frightening places I have ever visited!

I must mention a third cave before leaving Austria. The Salzberger Schacht is entered from the crest of the Untersberge mountain, fortunately accessible by cable-car. It is possible to ascend the mountain by this means late in the afternoon, descend the 435m-deep entrance shaft series to the cave beyond, explore this until four in the morning, and reascend in time for the first cable-car of the day. When *I* did it we mistimed, and arrived on the surface in a January blizzard an hour before the station opened!

Switzerland

Switzerland, better known for its mountaineering, is rapidly becoming a major caving area. It contains the Holloch, second in length only to the Kentucky caves and six times as deep. Another area, the Siebenhengste Hohgant, is likewise yielding long, deep and difficult systems.

Descending the 80 metre pitch, P26, Siebenhengste, Switzerland

Italy

Italy, with mountain limestone not only in its share of the Alps but also in the Apennines, has a strong caving tradition — and the caves to match. It is second only to France in the number of deep (over 500m) caves explored.

Spain

Spain saw the biggest increase of known caves of any country in the 1970s. Drawn by the warm air and cheap wine, cavers from Britain in particular thronged to Spain's Basque country. Two areas — around the village of Matienzo, and the Picos Mountains — have become happy hunting-grounds for those seeking long caves and deep caves respectively. (Please note, however, that foreign cavers wishing to go to Spain must join a group with an existing permission from the authorities, as new ones are not being granted.)

Dent de Crolles, France — Guiers Mort entrance

France

France, with major caves in the Alps, the Jura, the Pyrenees and elsewhere, has the strongest caving tradition of all. Edouard Martel, the first cave explorer systematically to survey his caves (and author of classic works on his discoveries in many countries in the period at the end of the last century), was the first in an unbroken line of 'spéléologues' combining physical daring, inventive genius and the ability to record discoveries in accurate detail.

The world's deepest-cave record has been held by a succession of French cave systems. At the end of the Second World War it was the Réseau du Dent de Crolles, in the Chartreuse mountains, north-west of Grenoble, at 623m. Although now eclipsed for depth, it is still a superb system to visit, with 35km of passages, and many fine through-trips possible between its five entrances. The Gouffre Berger, in the Vercors, just south of the Chartreuse mountains, was found and pushed in 1956 to a siphon (sump) at 1122m. Still very popular with cavers from all over the world, the Berger is not to be trifled with. Although a lot of its depth is gained in sloping passages, and the longest pitch is only 45m, it is very flood-prone, as several parties have found to their cost.

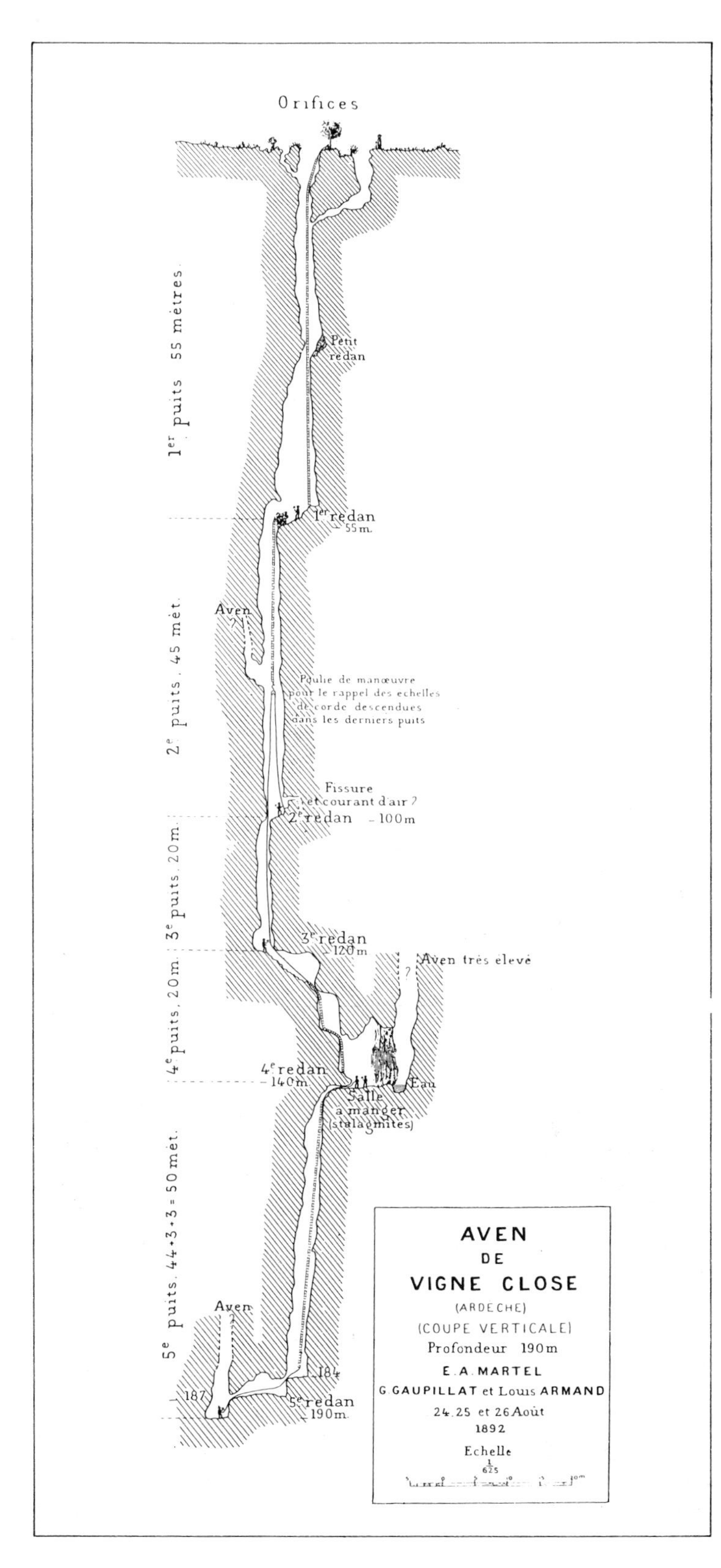

8th Cascade – La Cigalère, France

One of Martel's many surveys, illustrating not only fine draughtmanship, but also the amazing level of caving achieved with crude and heavy equipment nearly a century ago.

Meanwhile, in the Pyrenees, work has been going on in the Pierre St Martin system for over thirty years. The first entrance was a shaft of 320m, named the 'Lepineux' shaft after one of its discoverers. It was a formidable obstacle, and in 1952 disaster struck when one of the explorers, Loubens, fell to his death when the U-bolts on the winch being used broke. However, work continued, and what were then the largest known chambers in the world were discovered. There was also an underground river. The French electricity company, the EDF, decided to test this for hydroelectric potential, and drove a tunnel through the mountain into the biggest chamber, the Verna, which has a volume of four million cubic metres. The EDF could not harness the river, but did provide an easy route into the cave! Since then it has been explored to a length of the same order as that of the Dent de Crolles system by an association which co-ordinates the work of many visiting caving clubs — an organisation similar to, but more liberal than, the one which controls access and exploration of the Mammoth-Flint Ridge caves in the USA.

The Pierre St Martin overtook the Gouffre Berger for depth. In its turn it has now been overtaken by the Gouffre Jean Bernard. Tomorrow — who knows?

This quick world tour has been unfairly selective, and inevitably so. There is no space to record the many thousands of caves in every continent. I have missed out the 100km-plus gypsum caves in Russia's Ural mountains, the caves of Africa from Morocco to South Africa, the great open pits in Venezuela...

This is the golden age of cave discovery. Within a year or two of starting caving you too can tread where no one has gone before!

The Duck — Simpson's Cave, North Pennines, UK

Poulnagollum of the boats, Co. Fermanagh, Northern Ireland

Shannon Cave, Ireland, found 1982

Caves of the British Isles

British and Irish cavers have found caves virtually everywhere that limestone outcrops! Most of Britain's limestone is from the Carboniferous Period, laid down when the islands lay under a warm, shallow sea some 270 million years ago. This produced most of the caves, although a few are found in even older rocks (in Sutherland and South Devon).

Ireland is particularly rich in limestone — although a lot of it lies hidden under low-lying peat bogs. The most important caving area in Eire is County Clare, famous for the bare limestone scenery of the Burren. Beneath this run many kilometres of stream passages, fed by copious rainfall straight from the Atlantic. Some of these passages are of large dimensions, yet it is thought that they have formed mainly since the last major glaciation scoured and shaped the surface, around twenty thousand years ago. Further north, on the Eire/Ulster border,

Fermanagh has been known as an area of caves for centuries (the large river cave of Marble Arch was visited by Edouard Martel in 1895). Nowadays the area is known more for its vertical shaft entrances than horizontal ones, the Reyfad system being the most important.

England has three major caving regions, of which the northern Pennines, centred around the four peaks of Gragareth, Whernside, Ingleborough and Pen-y-Ghent, is by far the largest, with over a thousand known caves and potholes — so many, in fact, that six guidebooks are needed to describe them all! Because of the flat-lying, massive limestone beds, the caves consist of long horizontal passages linked by vertical shafts, so that although the deepest system (Pen-y-Ghent pot) descends only 184m, including the depth of a dive in the terminal sump, getting to the bottom involves long crawls and twelve pitches, accompanied by a stream which floods the system after heavy rain.

Gypsum Cavern — also part of the Lancaster Easegill System, UK

Main shaft, Gaping Gill, Northern Pennines, UK (100 metres)

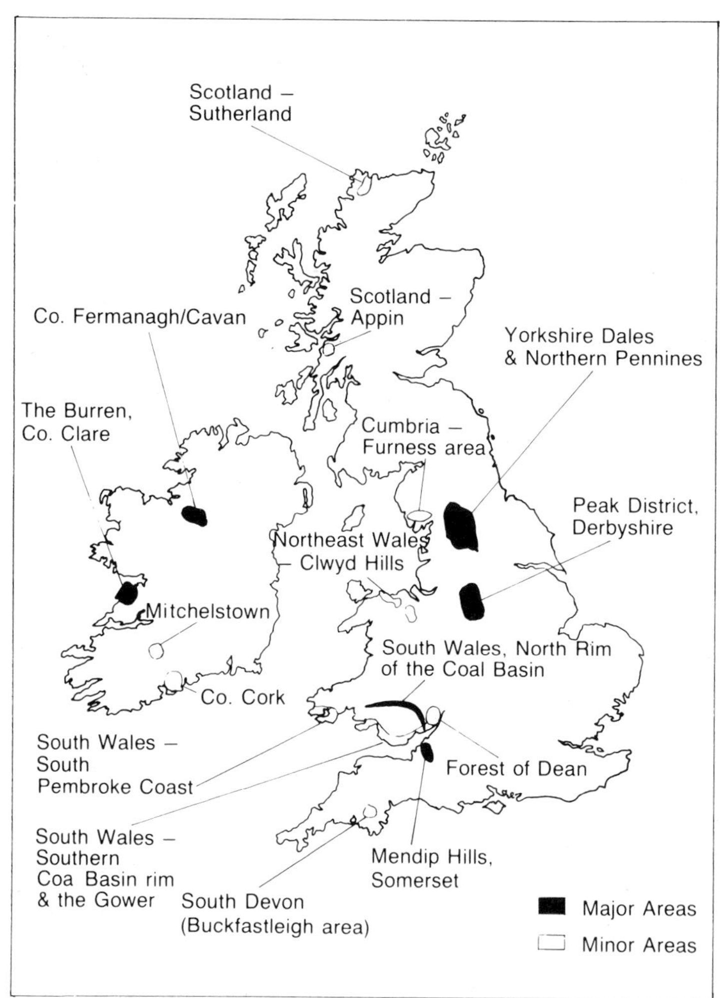

illustrated p.51
Lancaster Hole Streamway. Part of Britain's longest cave

Ever since their discovery in 1946–7, the caverns under Casterton Fell have been steadily extended. The last major breakthrough was in autumn 1978, when the Easegill-Lancaster Hole system was linked with Pippikin Pot, under the adjoining Leck Fell. The network now has more than 40km of known passage, making it Britain's longest. With so many caves it is difficult to choose only a few, but mention must be made of the 110m entrance shaft of Gaping Gill, on the flank of Ingleborough. The deepest shaft in Britain, it drops into the biggest chamber, from which an extensive network of passages leads off. At Spring Bank Holiday, and again in August, local clubs run 'winch meets' when, for a small fee, you can descend this great shaft in the lonely comfort of a bosun's chair!

Derbyshire has been a lead-mining area since Roman times, and miners have broken into many natural cavities in their search for minerals. Above the village of Castleton there are some particularly fine systems, some of which have show-cave sections.

A 'coffin level', chipped out of solid rock by miners in Knotlow Mine, Derbyshire, UK. So-called because of its shape

Peak Cavern, Derbyshire. Passage illustrating both phreatic and vadose development

The Giants-Oxlow system and the Speedwell-Peak Cavern complex contain interesting, and sometimes unusual, passages, partly due to being in reef limestone, rare for British caves.

The Mendip hills, in Somerset, contain interesting systems formed in steeply dipping limestone beds. Swildon's Hole is the longest, at 7km. Like other caves of this small area it is very heavily visited. There is so little accessible cave (about 25km in all) compared to the number of cavers based there that enormous energy has gone into digging for new caves, and into cave-diving. The resurgence of Wookey Hole has been dived through no fewer than 26 flooded sections, or 'sumps'.

Two major cave systems have been discovered in North Wales in recent years, and another in the Borders at Chepstow, but the really large stuff, and it *is* large, is in South Wales. Ogof Ffynnon Ddu is Britain's deepest cave, at 308m. Despite its depth it has few pitches, the extensive passages simply sloping with the dip of the rock at about 6 degrees. Nearly as long as the Lancaster Hole system in the north of England, it has a magnificent stream passage, and a complicated network of large fossil passages. Just across the Tawe valley lies Dan-yr-Ogof, another massive system of great interest containing many beautiful formations. In south-east Wales there are two huge systems under the gritstone cap of the Llangattock escarpment — Agen Allwedd and Ogof Craig-y-Ffynnon.

Scotland, famed for its mountain scenery, has no major caves, but determined work by local cavers has produced discoveries in Sutherland and Argyll. Best known is Smoo cave, which debouches onto the beach at Durness in the far north-west.

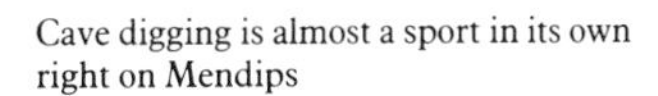

Cave digging is almost a sport in its own right on Mendips

Cascade, Swildon's Hole, Mendips, UK

The streamway, Ogof Ffynnon Ddu, South Wales

Caving internationally

This book can do no more than whet your appetite to learn more about individual caves. Getting enough detail to be able to locate caves is often difficult; and advance details about the caves themselves often impossible. Permission to descend caves can also be a problem.

British cavers are well served by detailed guide books which describe where to find caves, what to expect, and how to gain permission to go down them. Even so, the books go out of print all too frequently, so purchasing a set of guides should be an early event in every caver's career.

In other countries there are greatly differing attitudes to the publication of information on caves and the granting of permission to cave. There is a quite widespread view that cave guides should not be published, as they may lead irresponsible

people — or just too many people — to caves they would not find on their own. Where this is the case it is necessary to track down articles in obscure club journals and, if possible, make contact with cavers who know the area or country. In the UK the best source of information about caves in other countries is the British Cave Research Association. Its library and foreign recorder are the cavers' equivalent of the British Museum reading room! Whernside Cave and Fell Centre has another fine library.

Some countries regard caving as justifiable only if it can be classified as scientific research. Spain is one of these. Before being allowed in, a foreign caving group must establish its credentials and work closely with the local authorities. In other countries, local cavers are not too keen on foreigners coming in and stealing 'new' caves from under their noses, but are usually more amenable to plans to visit known systems, provided the local caving group is contacted first. This applies particularly in Eastern Europe.

Co-ordination of work on major caving areas can be a problem, especially if the area is one that attracts the attention of cavers from different countries. In the Pyrenees, the *Association de Recherches Spéléologiques Internationales à la Pierre St Martin* (ARSIP) directs teams of cavers from many countries to appropriate caving projects in its area. This is a much more open

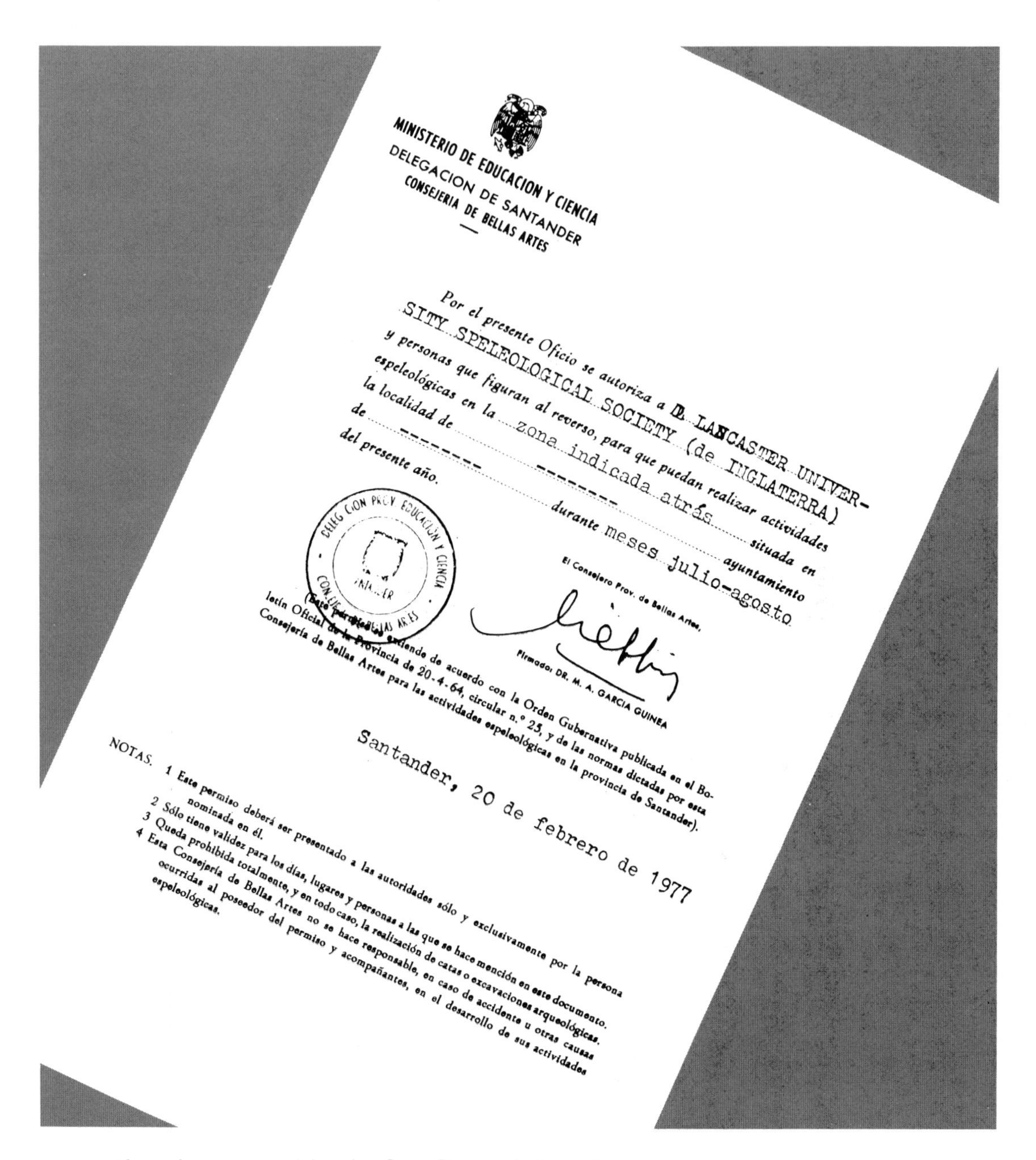

MINISTERIO DE EDUCACION Y CIENCIA
DELEGACION DE SANTANDER
CONSEJERIA DE BELLAS ARTES

Por el presente Oficio se autoriza a D. LANCASTER UNIVERSITY SPELEOLOGICAL SOCIETY (de INGLATERRA) *y personas que figuran al reverso, para que puedan realizar actividades espeleológicas en la* zona indicada atrás *situada en la localidad de* -------- *ayuntamiento de* -------- *durante* meses julio-agosto *del presente año.*

El Consejero Prov. de Bellas Artes,

Firmado: DR. M. A. GARCIA GUINEA

(Este permiso se extiende de acuerdo con la Orden Gubernativa publicada en el Boletín Oficial de la Provincia de 20-4-64, circular n.º 25, y de las normas dictadas por esta Consejería de Bellas Artes para las actividades espeleológicas en la provincia de Santander).

Santander, 20 de febrero de 1977

NOTAS.
1 Este permiso deberá ser presentado a las autoridades sólo y exclusivamente por la persona nominada en él.
2 Sólo tiene validez para los días, lugares y personas a las que se hace mención en este documento.
3 Queda prohibida totalmente, y en todo caso, la realización de catas o excavaciones arqueológicas.
4 Esta Consejería de Bellas Artes no se hace responsable, en caso de accidente u otras causas ocurridas al poseedor del permiso y acompañantes, en el desarrollo de sus actividades espeleológicas.

system than that operated by the Cave Research Foundation in controlling caving in the Mammoth Cave National Park. Before you can undertake exploratory caving there, it is necessary to be accepted as an individual member of the Foundation.

It would be wrong to finish this chapter in pessimistic vein. Caving must be one of the most international of activities. Caves are where you find them — and that is all over the world. There is a natural affinity between cavers and, although there may be rivalry between caving groups in the same country, international contacts are usually extremely helpful. The parochial caver (and there are some one-cave specialists!) is missing a lot.

Clothing and personal gear for cavers

Not long ago, there was no such thing as caving equipment. One just used whatever could be pressed into service. Edouard Martel, the great French pioneer, caved in suit and bowler hat, by the light of a candle, less than a century ago. Truth to tell, much caving *can* be done in ordinary clothing, and it is a mistake for the absolute beginner to spend a fortune on caving gear before he has found out whether this is the recreation for him. Do not be fooled into thinking that expertise can be purchased along with the equipment! This warning over, the rest of this chapter is mainly for the aspirant who, having got wet and muddy in his old clothes and enjoyed it, wants to do the job properly.

Potholing in the 1930s — tweed suits and ship's ladders!

Clothing

There is no such thing as the ideal, all-purpose caving outfit. Everything depends on the temperature of the cave and how wet it is. In tropical caves, with temperatures as high as 22°C (72°F: Jamaica) or even 26°C (79°F: Malaysia), shirt and shorts suffice, unless a thin boiler suit is preferred to guard against skin abrasion. However, most cavers live in cooler climes and need more protection, at least when under their home ground. Caves in the central USA have temperatures around 13°C (55°F: Mammoth Cave), and at this level hard-wearing denim clothing is ideal on dry trips. Further north, however, the temperature drops: caves in the UK, for instance, have a steady rock-temperature of around 8°C (46°F), with stream-water much colder than this in winter. In arctic and alpine regions, caves are often at or below freezing. (Less protection is needed below zero than just above — you cannot immerse yourself in solid ice!)

Clothing for these temperate and cold caves needs to protect the wearer against cold and wet, stand up to heavy abrasion, and allow the wearer to move easily. British cavers, with generally wet caves, were before 1980 almost uniformly dressed in wet-suits. Now, however, this is true only of cave divers, and most cavers in most countries prefer, on all but the wettest trips, a warm one-piece suit topped by a waterproof overall.

One-piece clothing has the advantage that it stays put in the middle, regardless of the contortions of the wearer. A one-piece undersuit needs to be a close fit, as its insulating properties are largely lost if it is flapping loose around the wearer. Suits of fibre-pile are favoured in the UK. Quality varies, with a good suit which offers adequate protection in cave temperatures

Tropical caving kit — Colin Boothroyd at top of a shaft in Clearwater Cave, Mulu, Sarawak. Suitable gear when the temperature is 26°C, but not for 'normal' caves

approaching zero costing about the same as a good anorak. 'Rexotherm' suits are more popular in mainland Europe. This material is a laminate, with thin fibre-mat and a metallised layer sandwiched between nylon or polyester cloth. The metallic layer is designed to curb radiative heat-loss. My own experience, based on cold bivouacs in Austrian caves, is that Rexotherm suits are fine for active caving, but fibre-pile suits keep you much warmer when you are merely standing around.

To keep bone dry in a fibre-pile furry-suit requires a totally waterproof layer over it. This is necessary only if long stretches of deep water are to be negotiated, or if the water is very cold. Ex-service garments, made as 'once-only' suits for use in case of shipwreck, are sporadically available for a few pounds; but a French firm, Gomex, makes dry-suits for the job. Best described as flexible rubber body johnnies, usually in shocking pink, they must be the ultimate for rubber fetishists. Despite their dubious appearance, they work well, but great care must be taken not to puncture them or to allow water in through the neck-hole. A dry-suit full of water is a dangerous liability.

Petzl caving oversuit

Whether you are wearing just an undersuit or undersuit plus dry-suit, an oversuit, preferably waterproof, should normally be added. Its functions are to deflect as much water as possible, guard against heat-loss by evaporation, and prevent clothes underneath from wearing out. The best ones are made of PVC-impregnated nylon or polyester. They are very tough, but can be glued and patched if necessary. An integral hood of thinner, flexible material which can be worn under the helmet is a good feature to look for. Some cavers like elasticated cuffs, but these can chafe if worn for long periods; and plain ends have the advantage of allowing you to make watertight seals with gloves and with wellies by folding the suit ends carefully over and securing them with rubber bands cut from inner tubes.

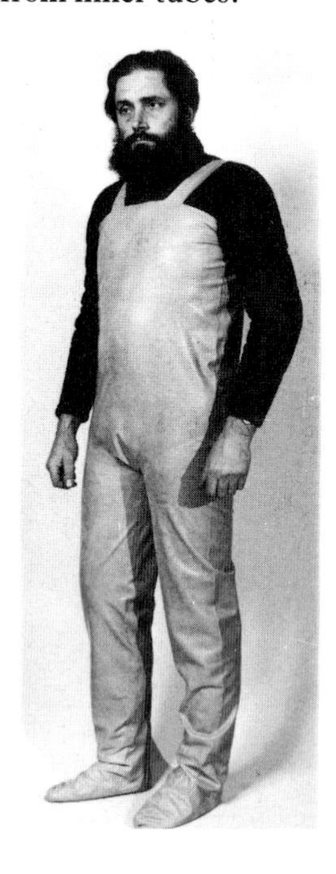

Fibre-pile caving undersuit

Gomex dry-suit or pontonniere

Most cavers will get along well with an under- and over-suit for general caving, adding a dry-suit later. However, in some areas wet-suits are necessary or may be preferred. A wet-suit is a tight-fitting suit of expanded closed-cell neoprene which acts as an extra layer of insulation on top of the skin. Water does penetrate between skin and suit — hence the name — but, once there, the thin water layer stays put and soon warms up. Minor tears in the suit are not of major importance — a very useful feature underground. Wet-suits are, of course, used in many other sports, but if you are buying one for caving the following features are desirable:

Caving wet-suit

- Suit thickness of 4mm for most caves and water, but 6mm for caves below about 7°C (45°F).
- Padded knees and elbows. (Remember that you crawl on the patch *below* the kneecap.)
- Always choose nylon-lined neoprene. Unlined material rips too easily.
- A one-piece suit is best if for caving use only.
- The front zip should be shielded from abrasion by a flap of neoprene.
- Last, but not least, a good fit is essential. Too tight, and you will be chafed, develop IBS (Itchy Bum Syndrome), lose arm strength due to constricted circulation, and feel as if you are wearing a strait-jacket. Too loose and the garment does not work. A wet-suit is at its best when the wearer is immersed, giving warmth and buoyancy, but it does not provide particularly good insulation if you are hanging around for long in cold, draughty passages.

Cave divers need more sophisticated wet-suits, together with hood, gloves and neoprene socks, but space does not permit inclusion of details. However, gloves of thin (2–3mm) neoprene are useful for 'ordinary' cavers in cold conditions, and neoprene socks are warmer than soggy wool — they also dry out overnight!

Expanded neoprene makes the most comfortable knee-pads. These are advisable in any cave involving extended crawling. Although miner's knee-pads are more robust, it is difficult to do anything other than crawl in them, as they become most uncomfortable when you are standing up.

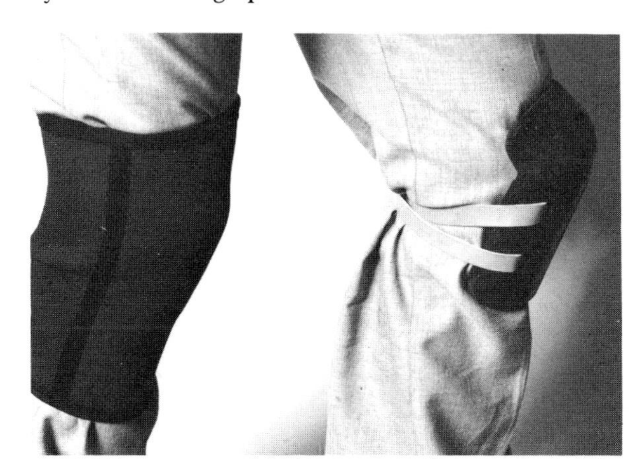

Left leg: neoprene knee pad. Moderate protection but comfortable to wear
Right leg: miner's knee pad. Excellent when kneeling but awkward at other times

Boots

Always wear boots underground in anything more difficult than a show-cave: apart from the need to grip rock and mud, it is very easy to bang your ankles in the dark. Wet caves make short work of leather boots, and all-synthetic wellies or ankle boots are better value. They should have a well cleated sole, and there is considerable evidence that a relatively soft sole compound grips better than a hard one.

Good fit is obviously important. Neoprene socks help in this respect, especially with wellies. The way in which a waterproof seal can be made between oversuit and wellies has already been explained (see page 58).

Avoid boots with snags. Hooks for lacing provide the prime example of what to avoid, but blobby toes caused by front-end protection can also be counterproductive.

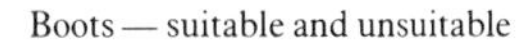

Boots — suitable and unsuitable

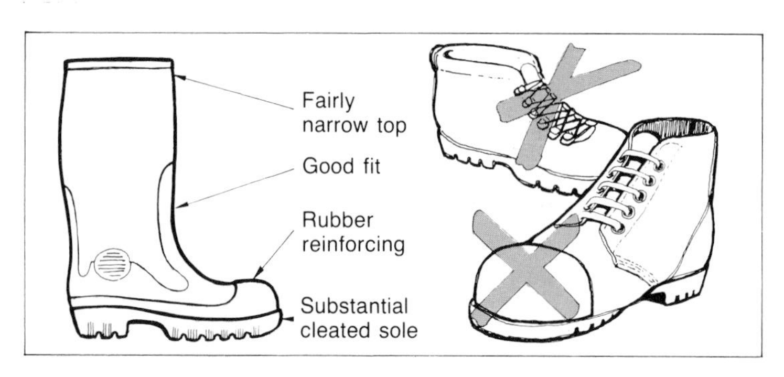

Headgear

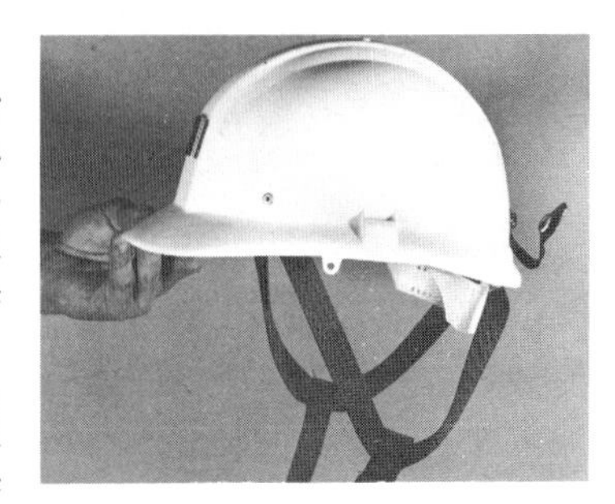

A basic helmet with chinstrap and lamp-bracket

Foot protection is important, but head protection is vital. For horizontal caving, helmets to the specification used by miners suffice. These relatively cheap and lightweight helmets are designed to hold a headlamp in place and protect the wearer's head against bumps and scrapes. They are not adequate to protect the head in event of a major fall.

For potholing, a helmet to climbing specification is better, although more expensive. Whichever you use, it must be a good fit to prevent the brim flopping down onto your eyes with the weight of the lamp above, and have a good Y-type chin-strap to prevent it being knocked off.

Beware of cheap and nasty building-site helmets. Many have pop-in headcradles which allow the helmet shell to 'pop out', leaving the wearer with a silly bit of plastic strip on his head.

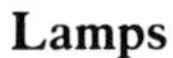

Lamps

Clothing is clearly important to the caver, but of paramount importance is effective light. You *can* 'streak' through a cave with a light, but even superbly clothed you will not get far in the dark! Main light for the caver comes from either a battery lamp or an acetylene-gas lamp.

Carbide lamps

Acetylene lamps, more commonly referred to as carbide lamps after the fuel they consume, come in small helmet-mounted models and larger versions with a burner on the helmet and a gas generator clipped to the waist. We will examine the small ones first.

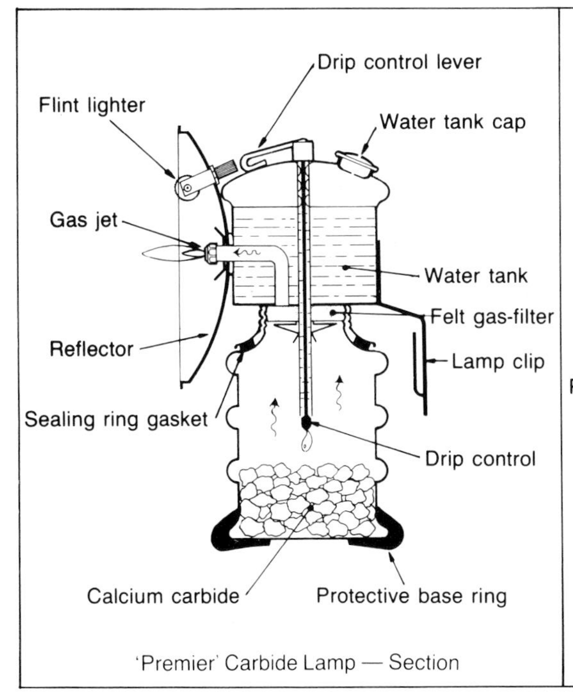

'Premier' Carbide Lamp — Section

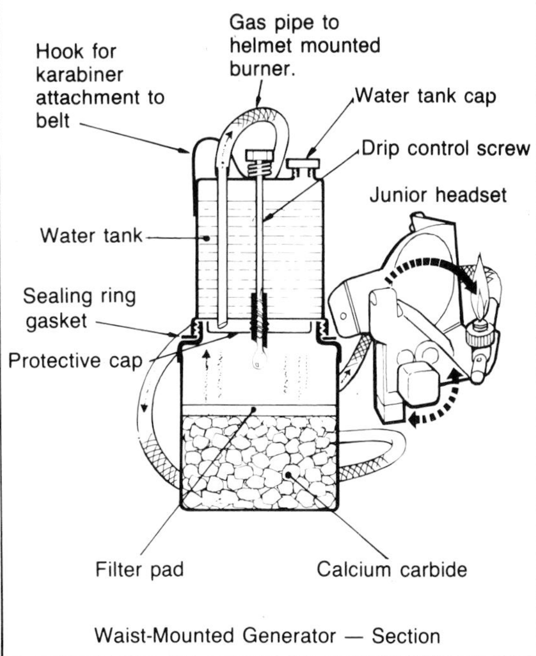

Waist-Mounted Generator — Section

Carbide lamp in action, under the Picos mountains, Northern Spain

'Premier' carbide cap-lamp

Plastic lamps have a very poor reputation among cavers. Far better are the brass lamps made by Premier. One filling of calcium carbide, small size (do not fill the container much more than half full), will last three to four hours.

How to use the lamp:
Fill water tank; fill carbide chamber three-fifths full; screw both tightly together. (NB: do not lose the rubber washer.) Open the water dropper three to four notches only. Wait a minute; cup hand over reflector; rub palm smartly across flint wheel. This should ignite the lamp, which will take ten minutes or so to settle down to a steady 20–30mm flame.

Care of the lamp: Make sure it is completely clean before use. Clean the jet carefully with a jet pricker. *Empty spent carbide into a polythene bottle or rubber bag and put it in the dustbin back home. NEVER leave it in or near a cave.* (This means, of course, that you need *two* waterproof containers — one to carry your spare carbide into the cave, and another to carry the spent fuel out.)

Small carbide lamps, often known as 'stinkies', are cheap, and, if well maintained, quite reliable. However, their weight tends to pull your helmet down over your eyes, they need frequent refilling (with attendant problems of spent and unspent carbide), and pose a pollution threat unless used carefully. Some of these difficulties can be avoided by using a large lamp.

A bigger acetylene generator on the helmet would be quite impractical, so the larger lamps are slit in two, with a gas generator carried at the waist being connected by tubing to a jet mounted on the helmet, usually accompanied by a piezo electric ignitor. The normal size of waist-mounted generator takes 200g of carbide (use lumps of 10–20mm size) per 'fill', and this makes enough gas for eight hours of really bright light, using a jet larger than that in the small lamp described above. The best performance is gained, as before, by not overfilling, and by putting the 'fill' into a thick nylon stocking bag, which acts as an extra filter, preventing the lamp from 'gunging up'.

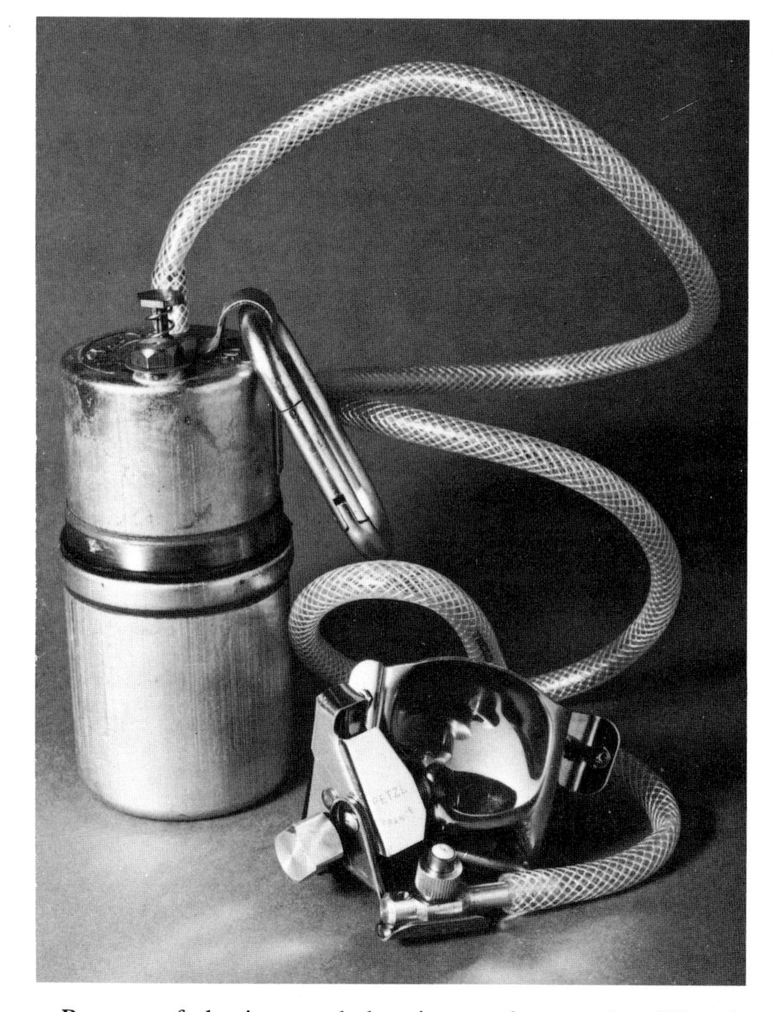

Acetylene generator, carried at the waist, and Petzl headset

Because of the increased duration, underground refilling is unnecessary on most trips. However, the greatest advantage of all carbide lamps is that their use can be extended almost indefinitely by carrying sufficient fuel. This makes them standard lighting for all extended underground trips, and for expeditions away from electrical recharging facilities. Small amounts of carbide can be carried in wide-topped polythene bottles with a waterproof seal, but for long trips and larger amounts, a home-made container

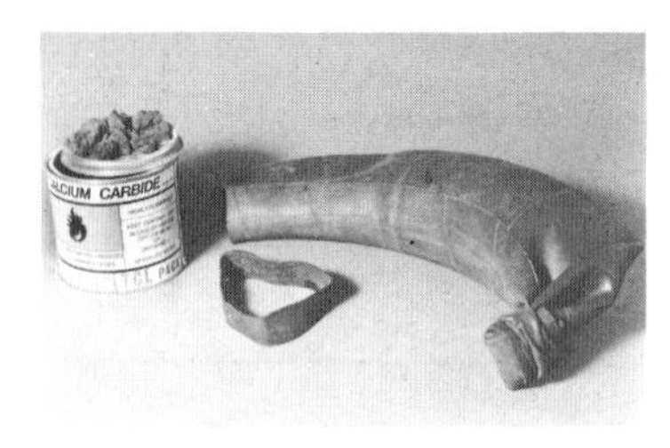

Carbide carrier made of inner tube

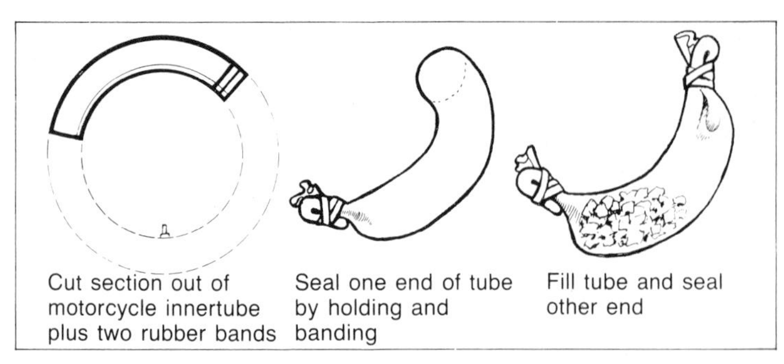

may be better. Obtain an old motorcycle inner tube, cut out a section about 0.6m long, and some short bands. Fold one end of the tube twice, and secure it with a band. The resulting container can be filled with calcium carbide, and sealed by double-folding and banding the other end.

Cut section out of motorcycle innertube plus two rubber bands
Seal one end of tube by holding and banding
Fill tube and seal other end

Never carry used carbide in a sealed metal container. If it is still giving off gas, the container may explode.

A full expedition lighting set will have an auxiliary dry-battery electric lamp, for use in wet conditions (carbide lamps go out underwater) and when refuelling. A good example is the Petzl mixed light set. This has a battery case mounted on the back of the helmet, counter-balancing the headlamp. Since the electrical component will be used only intermittently, it is worth investing in alkaline dry-batteries, such as Mallory Duracell, as 'ordinary' batteries deteriorate in a few months even if not used.

Expedition type lightset, with an acetylene flame and auxiliary electric lamp

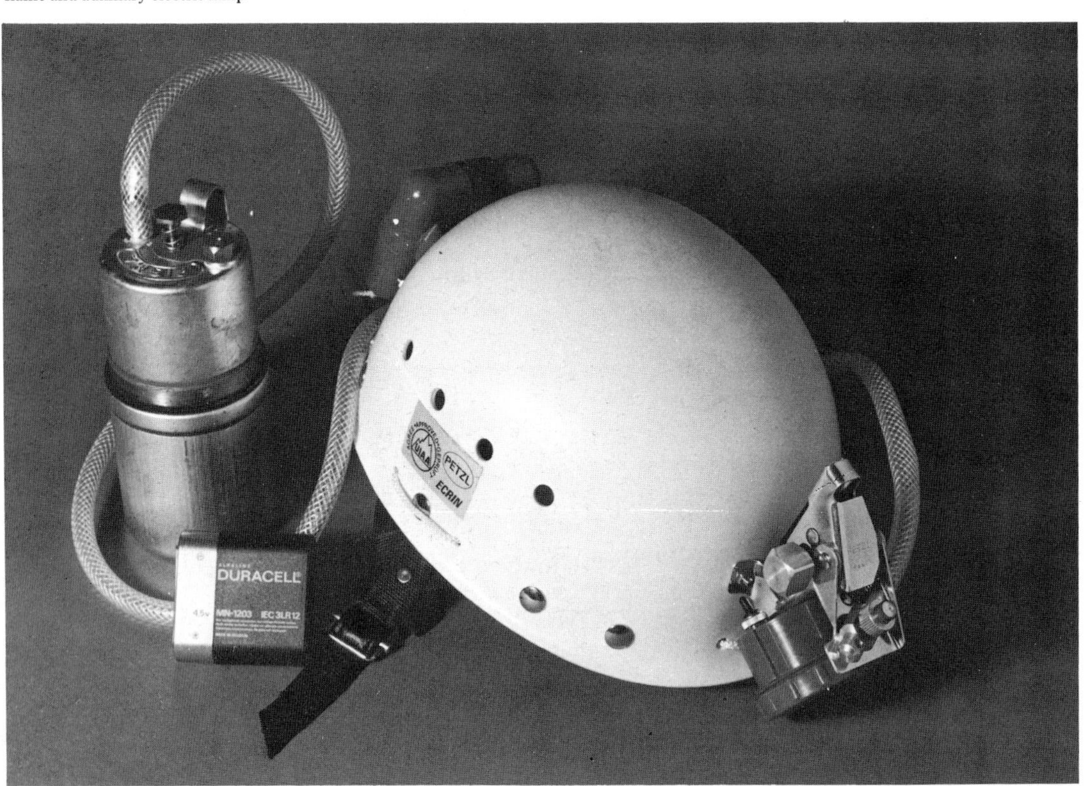

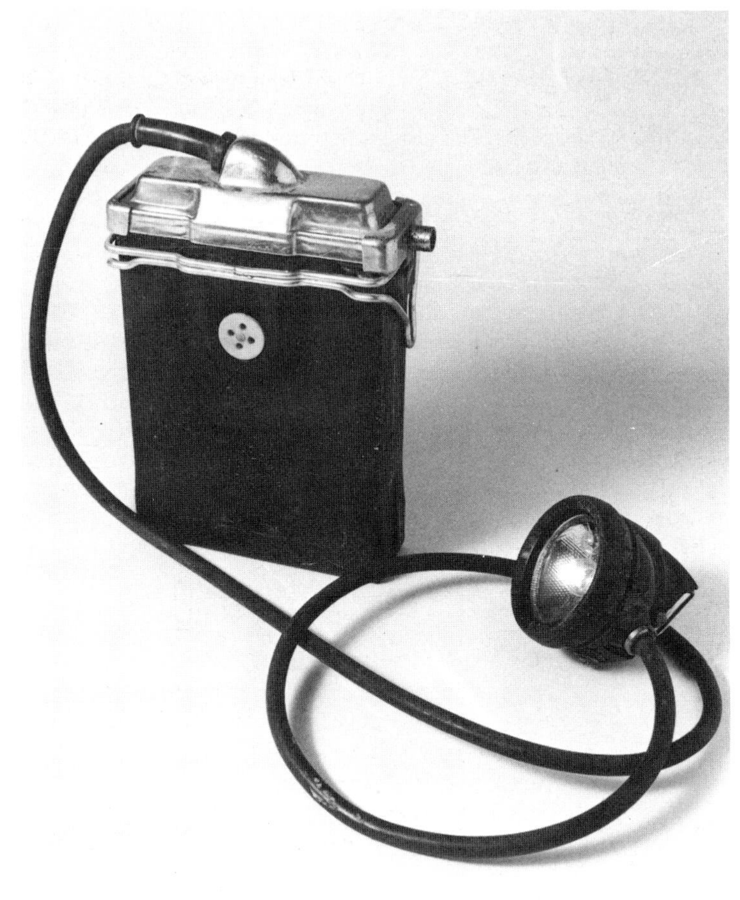

Rechargeable lead-acid accumulator and headlamp

Electric lighting

In the UK, with its short but wet caves, electric lighting is preferred as the main light. Most of the head-torches on the market — and it must be a head-torch — are so flimsy as to be useless for caving. The Petzl version, made of nylon, is robust, but does not give enough light for a main beam, and except on short, easy excursions, should be used only as a reserve.

For a main electric lamp it is best to ignore throw-away battery lamps, and go for a rechargeable one. Those developed for mining use, such as the Oldham, have been standard for many years. Like the big carbide lamp, they are in two parts, with an accumulator, weighing over 2kg, carried on a waist belt, and a headlamp. Obviously, the two are connected by a cable.

The two main types of miner's-lamp accumulators contain sulphuric acid and potassium hydroxide (alkaline), respectively. Both chemicals can be very nasty if spillage occurs, but the effect of the alkali is much worse on the skin — so much so that it is best to steer clear of such lamps unless you really know what you are doing. If you buy a miner's lamp, ask for a lead-acid model and charger. Nevertheless, beware of acid spillage onto any equipment, particularly nylon ropes and harness. Most lead-acid lamps are designed to recharge through special contacts on the head-piece, and the charger should be fitted for this.

Charging key for the 'Oldham' headlamp

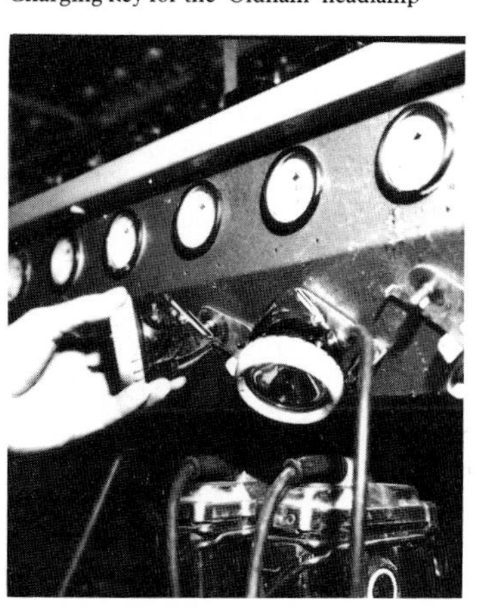

These lamps are made to give good light for a full eight-hour shift underground, and will give even longer on main-beam at first. They have a second, pilot bulb, which gives less light but for twice as long. If a long trip is in prospect, make as much use of pilot bulbs as possible, switching off completely when resting. In this way an electric lamp can be stretched to 'do' quite long trips. With care, the accumulator will last three years, and a replacement then will cost only half the price of a complete new lamp.

A simple battery belt, and a load-bearing model to which a rope can be attached with a karabiner

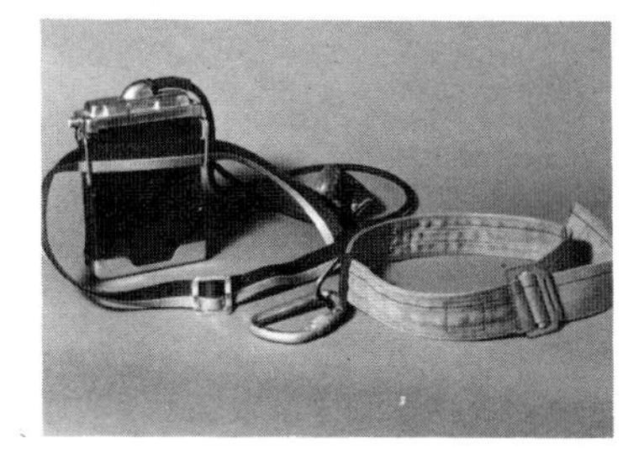

Things are changing, though, and the disadvantages of a potentially leaky lamp fitted with seals that prevent dangerous tampering in coal mines, but are a pain in the neck to the caver, have been recognised. Rechargeable closed-cell lamps specifically made for caving are starting to appear. As yet they have enjoyed only a fraction of the development of miners' lamps, but they can only get better. Removal of the threat of electrolyte spillage makes them much safer,and they may prove to have a longer life. Hence they should be seriously considered by the new caver. Speleotechnics have produced an extremely robust sealed-cell rechargeable caver's lamp in the UK.

'Speleotechnics' sealed-cell rechargeable lamp

Belts

Whether you use a rechargeable electric lamp or a carbide generator, you will need to wear a belt to hold it. This may be a thin polypropylene-tape belt with a simple buckle, or, for a few pounds more, a load-bearing belt with a locking buckle (see above). In the case of a load-bearing belt, if the accumulator to be carried on it is lead-acid, it *must* be of polyester; if alkaline, of nylon. The load-bearing belt doubles as a belay belt, and if it is used as such should have a karabiner, or snap-link, clipped to it. For this purpose it is best to use a 12mm offset 'D' screw-gate model, in steel or alloy.

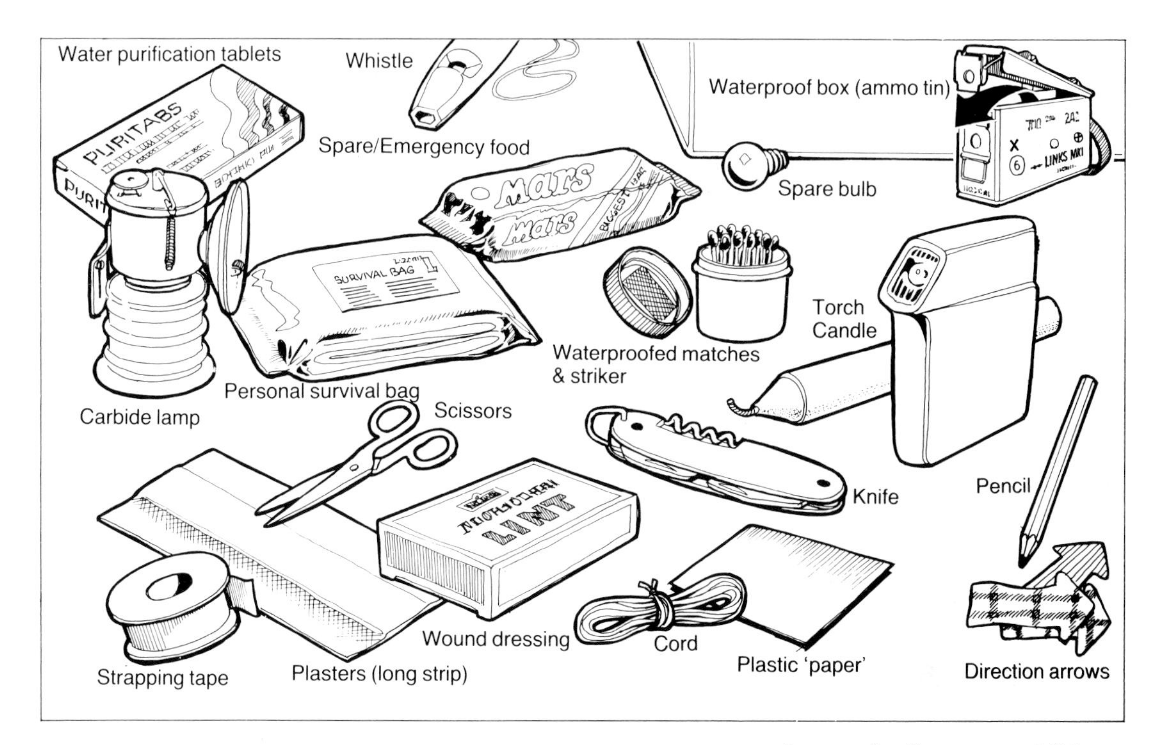

Be prepared! — You may not need all the items shown here on every trip, but you should always consider what you might need, and make sure you carry those things

Emergency items

If all goes well you will need no more than clothes, boots, helmet and lamp for your caving trip. However, a few emergency items can be very useful from time to time, and these are listed below. You will not need all of them for every trip, and sometimes one item will suffice for a whole party. You must use your intelligence!

- **Light spares:** Spare bulb, fuse, jet-pricker.
- **Spare light:** Small carbide lamp or torch in case main light fails. Candle and matches in a waterproof box for emergencies. 'Chemilight' stick. (NB: has limited shelf-life.)
- **Food:** Emergency rations for unexpected delays, in waterproof wrapper. Avoid squashy items.
- **First aid:** Thin polythene body bag to prevent exposure if trapped or lost. A lifesaver that will fit in your helmet. Plasters, elastic strapping, dressings, scissors, whistle, etc.
- **General:** Watch, knife, string, survey of your cave.
- **And:** An ammo tin or tackle bag to carry it all in!

Learning to cave

This book is no substitute for practical experience. It will give you a good idea of what caving is all about, the gear you will need, and some information on where to find caves, but to use all this information safely requires great care. You can visit show-caves on your own, but doing the same thing in caves in their natural state requires great caution. Holes in the ground may be simple and safe, or they may equally be death-traps. I should hate this book to add to the newspaper headlines of the type: *'Rescuers risk lives for inexperienced cavers'*.

The best way to sample the delights of the underworld is by going on a course at a reputable outdoor centre. This will give you a taste of caving, following which you can join a club or decide that reading books about it is quite enough. The UK has its own national cave training centre at Whernside Manor, in the northern Pennines, and in France the *Fédération Française de la Spéléologie* has a similar centre at St Martin in the Vercors. In other countries it is not so easy, and it may be necessary to approach a club right from the start.

Club traditions in caving are very strong, and it is not always easy to be accepted into what may well be as much a social as a sporting association. Play the field — if there is one. Get a list of caving clubs from the National Caving Association, and write to the secretaries of possible ones. Some clubs are based where their members live, while others have a 'hut' in a particular caving region and may have members hailing from many places.

A third type is the club associated with a college or university. A very high proportion of young people start their caving with such clubs. Some are excellent, full of enthusiastic and talented cavers. However, their safety record is not good, probably because of the rapid turnover of members and the resulting lack of continuity of experience. My strong advice to anyone thinking of

Some club huts are next to caves. The Craven Pothole Club has a different neighbour

First descent. A *Guardian* reporter finding out what caving is about in Yordas Cave, Northern Pennines, UK

starting caving at college is to gain basic skills on a caving-centre course first. The college club will be delighted: it will show that you are really interested, and save them some initial training at a time when they are swamped with 'freshers'.

If several of you want to start caving you will learn a lot by progressing as a group, starting with easy caves and tackling more difficult ones as you learn — you may find this preferable to being the only novice in an experienced party. However, if you do take this option it *must* be under the guidance of a competent and experienced caver, capable of checking your progress, and able to suggest which cave system you should try next. I cannot stress too strongly how important this expert guidance is. The combination of inexperience and over-confidence results in more accidents underground than anything else. Techniques, especially, need to be properly taught and learned. Because of the nature of caving you may see little of other groups and their methods, and can make many dangerous mistakes unless you are being taught by someone who really knows.

So, get proper training, join a club, and good caving!

Getting about in caves

Narrow passages

Most caving does not need special skills, just the agility and fitness to move around in some rather odd ways. Cave walls contort to push you into all sorts of shapes, with water or mud often present for lubrication. Your first trip underground will probably leave you aching in every muscle! At times you may feel that you are reverting to infancy, spending more time on your hands and knees, on your stomach, or just squirming about, than on your feet.

Passages around one metre high can sometimes feel worst of all. Not low enough to demand all-out crawling, but too low to walk along properly, they demand a sort of shuffling crouch, with knees squashed up to your chest. In this position it is sometimes more comfortable to walk sideways like a crab than directly forwards, turning your head to see in front.

Crawling itself takes less effort, but it is slower and can be very hard on the knees. Girls and women seem particularly prone to bruising, and may find some sort of knee-padding desirable in 'crawly' caves — indeed, in the seemingly never-ending crawls of the Kentucky caves, it is considered essential, and guides refuse to take people underground without pads.

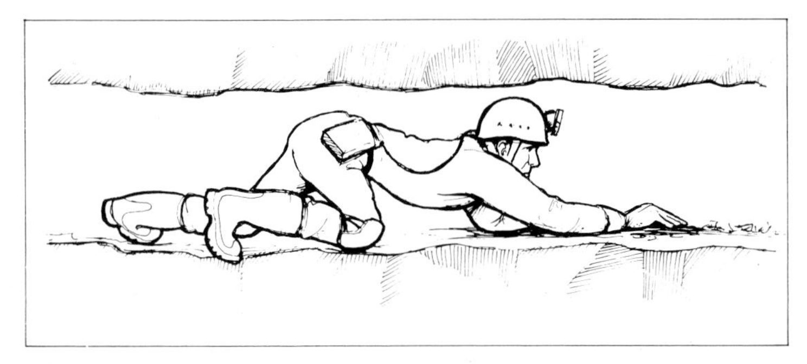

Stooping passages can be very tiring (dwarfish stature helps!)

illustrated p.71
Negotiating a connecting hole in Ogof Ffynnon Ddu, South Wales

Flat-out squirming through 'squeezes' requires more technique. Horizontal ones are not as bad as those you have to go through sideways. Make sure you are not going to be snagged up by a battery case or anything else before getting into the tight bit (you can push gear in front of you through the hole). Keep your arms in front of you, so that they cannot jam between body and wall. Push with your feet, squirm with your stomach, and use hands and elbows to ease yourself forwards.

In men the biggest part of the body is usually the chest. By breathing out, moving, and breathing again while resting, very tight squeezes can be negotiated. For women the problem is rather different. Despite appearances, it is not at chest level that they tend to stick: breasts compress more easily than buttocks!

Your size and motivation will determine the tightness of gap that you can squeeze through. Like being a jockey, caving is one

Wet-suited caver negotiating a squeeze in Ogof Hesp Alyn, North Wales

of the few physical activities where being small is a positive advantage. Thin people weighing around 60kg can usually squeeze through a 'letterbox' 180mm high; if your weight is nearer 80kg your limit is likely to be nearer 225mm.

Boulder chokes — places where the solid rock has collapsed to give a jumble of boulders blocking the passage — need special care. All sorts of movements may be needed to contort your way through the labyrinth cleared by the cavers who opened up the choke. Some have boulders which are cemented together by calcite deposition, and are relatively safe; others are being dissolved away by running water, and may be very unstable. The best way of moving through such a boulder choke is very carefully, going slowly and not pulling or pushing on anything that looks loose, with only one caver at a time in the loose bit.

Boulder choke

Larger passages

Not all cave passages are small: some are very large. Tall, thin canyon passages, caused by the downcutting action of cave streams, are particularly common. It may be necessary to climb, descend or move along these rift passages at varying heights. By using both walls and a little skill, this is usually quite easy.

'Bridging' is the best technique. By straddling the passage and pushing outwards with hands and feet, you find that minor bumps and ledges, including very sloping ones, become perfectly safe, and with practice provide an easy means of progression. If the passage gets too wide to be straddled, you may have to keep both feet on one wall and lean forwards with both hands on the opposite wall. This is more alarming and less secure. On very sloping holds it works only if you are semi-horizontal, thus exerting more sideways grip on the holds.

Jeff Clegg on the traverse en-route to Ogof Ffynnon Ddu III, South Wales

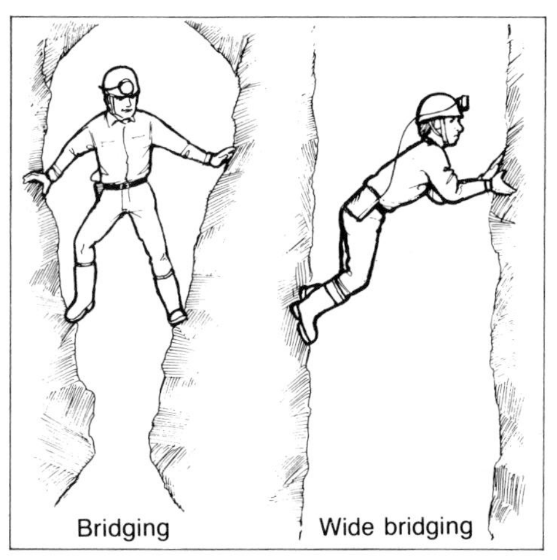

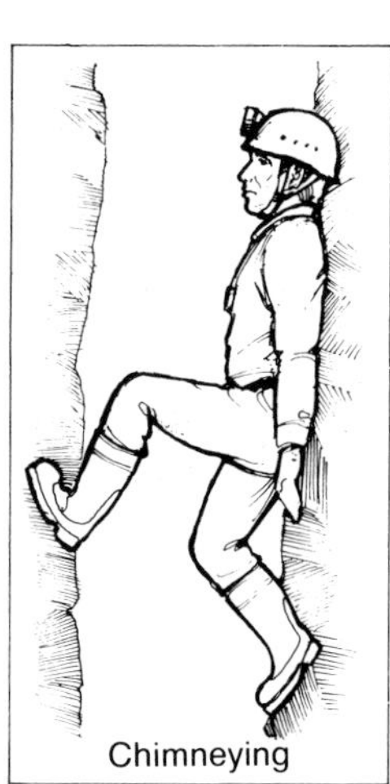

If the rift is too tight to bridge, 'chimneying' is necessary. Bracing yourself sideways in the rift, you wedge knees, feet, elbows, hands and back as necessary, and move — strenuously!

These techniques are safer than the normal mountaineer's practice of ascending a single face, which should be adopted only if there is no other way. Fortunately for cavers, most rock underground has good friction, even when wet. This surprises walkers, used to slipping around in the rain, but the reason for it is simple. Wet algae covering surface rocks make them slippy, but without light there are no algae, and hence no problem! Conversely, though, there is no vegetation to bind muddy or loose slopes: wet sloping mud can be extremely hazardous, so make sure your boots have a sole which will grip on it.

If there is *any* chance of falling, the 'move' should always be protected. The simplest form of safeguard is the handline. A rope is belayed at the top of the drop, which is descended with feet against the rock and hands on the rope, leaning out all the way. Odd though it may seem, the more you lean out on your arms, the less chance there is of your feet slipping. Some handlines are left in place in caves (they may be of chain rather than rope). They, as well as all other fixed aids, should be treated with great caution.

Caving can be muddy!
Apart from making you look like this it can also be a hazard in sloping places

Chimneying

Kingsdale Master Cave, Northern Pennines, UK

Accidents have occurred through cavers trusting old fixed aids in caves.

Horizontal traverse lines are used in a similar way. The rope — or wire — is fixed at both ends and used as an aid to traversing above a deep pool or a drop. If the traverse is long, or there is any danger of slipping, it is best to clip a 'cow's tail' into the rope with a karabiner (see illustration). Two cow's tails are needed if there is an intermediate belay along the traverse; then the second one can be clipped to the next section of the traverse rope before the first is unclipped.

Handline
Lean out as near horizontal as possible to prevent feet slipping
SLIDE hands up or down rope so that both are on it all the time

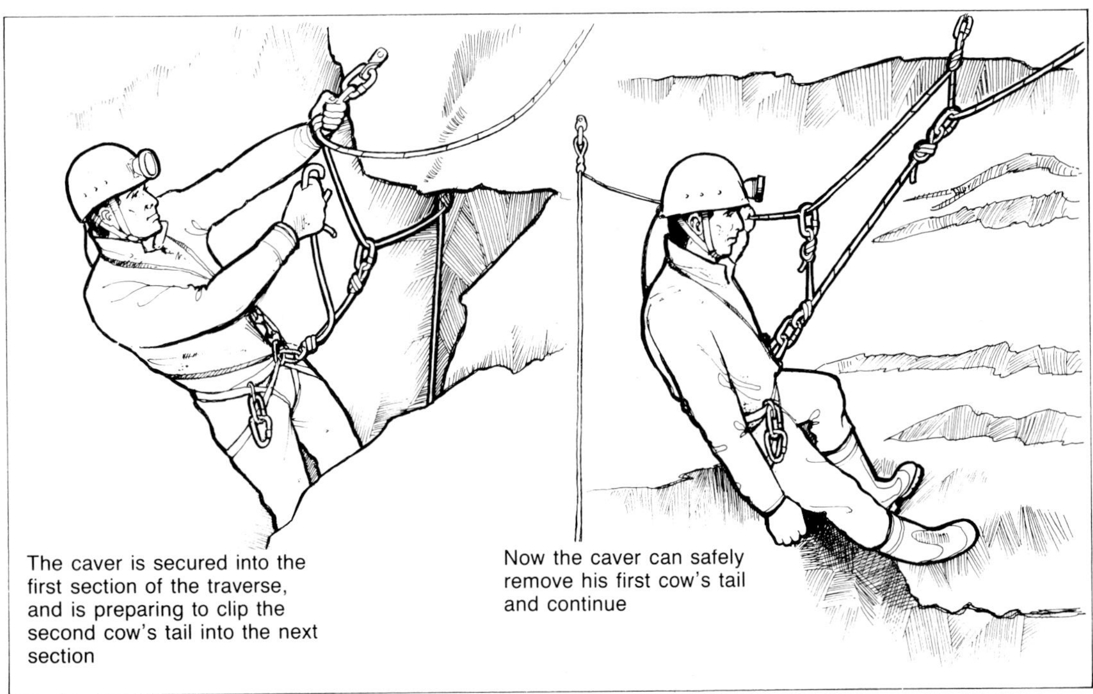

The caver is secured into the first section of the traverse, and is preparing to clip the second cow's tail into the next section

Now the caver can safely remove his first cow's tail and continue

Water

Water in caves always demands extra caution, with protective clothing if cold (see pages 120-126), and precautions against drowning whatever the temperature. If swimming is unavoidable, make sure that every member of the party is made buoyant by means of a wet-suit, a lifejacket, or a handline. Even strong swimmers have been dragged under and drowned by the weight of their equipment, and deep water may have dangerous currents.

Occasionally, a rubber dinghy will be the best way of crossing a short section of deep water. Unless it can be pulled backwards and forwards on a rope from each side of the water, it must be big enough to hold at least two cavers, so that all the party can be ferried across. Great care is needed to avoid punctures on sharp limestone. Table-tennis bats are usually more practical than normal paddles.

The entrance lake, Grotte Gournier, France

Floatation is essential in deep water

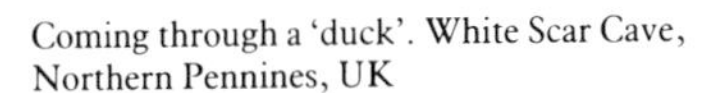

Coming through a 'duck'. White Scar Cave, Northern Pennines, UK

Caver emerging from a free-dive sump. Sump One, Swildon's Hole, Mendips, UK

Sometimes the water in a cave passage comes nearly to the roof. This is called a 'duck', and needs careful negotiation. You may have to turn your head on one side to be able to breathe, or even lie on your back with only your nostrils out of the water! If the duck is short it is often simplest just to hold your breath. Whichever method you use, be careful. This is not a place to panic.

Passages filled right to the roof with water are called 'sumps'. Very short ones can be free-dived, using just a lungful of air. However, should anything go wrong you are liable to drown, so great caution is necessary. The two basic rules are:

- Never attempt to free-dive an unknown sump.
- Always have someone in the party who has 'done' the sump before, and can guide the others.

There is usually a guideline to help you; this can also be used to signal, by means of tugs. Although some cavers free-dive sumps of ten metres or more, you are well advised to gain considerable experience before venturing into anything more difficult than well frequented sumps of two metres or less.

Cave diving using compressed air is almost a sport on its own, and beyond the scope of this book. If you fancy the idea, become a proficient caver *first*. Divers who have tried their hand in underground waters before knowing about caves have a very bad accident record. Cave diving is, however, becoming the most effective way of discovering new cave passages, especially in well explored regions, and is an increasingly important branch of caving.

From all this, you will have realised that the ideal caver is a midget contortionist equipped with gills! Do not despair: ordinary people who can stand a bit of wet, cold and mud manage just fine.

Martyn Farr all kitted up for a cave dive

Potholing techniques and equipment

Equipment

The evolution of equipment and techniques for vertical caving has been very rapid, and is still proceeding. Inevitably, this progress has been resisted by those who are quite happy with the old ways, but that is no reason for the beginner to start by learning anything other than the most modern ways. They are based on having a fixed rope down pitches and along traverses: the caver then clips himself to the rope so that he can travel along, down and up it in safety.

If we are to make sense of the techniques, we must first look at the equipment.

The rope

Fixed ropes require strength, abrasion resistance, shock-absorbency of about half the level needed in climbing ropes, and low stretch; in addition, non-twist construction makes ropes easier and safer to use.

Types of caving rope. Left to right 1. Beal Dynastat (Low stretch core, shock-absorbing outer sheath) 2. Edelrid speleocord (ropelets inside sheath) 3. Blue Water II (ropelets inside sheath)

The rope cleaning machine at Whernside Cave & Fell Centre.
Fixing the rope in position. Note brushes and water supply pipe.
With the door shut, the rope is drawn three times through the sets of lubricated brushes

Speleo ropes are made to these specifications. Most are of nylon construction, since only this polymer has the right combination of shock-absorbency and strength, although there have been some very recent developments of composite ropes, with polyester or kevlar cores and nylon outer sheaths, which may alter future thinking. There are many makes of rope on the market, varying in diameter from 9.5mm to 11mm, with some thinner 'half ropes' suitable only for small and very experienced parties. A few makes to look out for are Blue Water (US), Edelrid (German) and Beal (French).

Buying a good rope is the easiest part: it is proper care and maintenance that will keep you safe. Soak and dry your rope before measuring and cutting it to length for actual use: it will shrink by 10 per cent or so and tighten up, keeping grit out of the core. Avoid contact with chemicals: acid rots nylon, and alkali affects terylene. Becoming neurotic about the possible effect of battery-acid spillage on ropes is a good way to stay alive. Clean your rope after each use, and inspect it carefully for any damage.

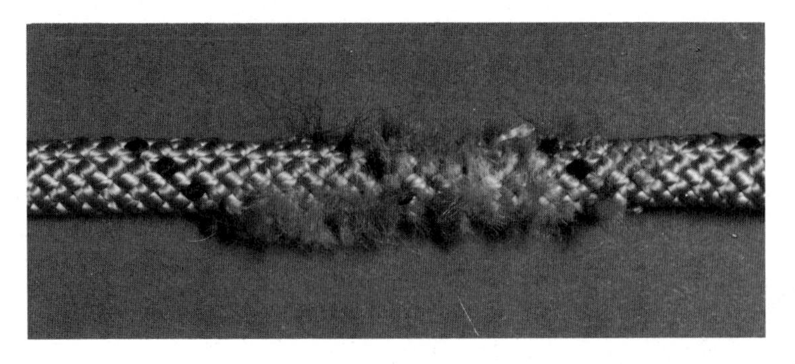

Damage caused by localised rubbing. The rope should be cut here, and the two halves then used for shorter pitches

Running it slowly through your hands and looking at it carefully at the same time is the best way. Cut it in two if you spot an abrasion point, and use the bits for shorter pitches. Do not dry ropes, or leave them out, in sunlight as ultraviolet light degrades them. Store them in a cool, dark place. It does not matter if it takes them some time to dry out.

In use, ropes should never be trodden on or dragged through mud, and should be protected from abrasion by hanging them where they do not rub against the rock. If this is impossible, rope pads should be used, although they work only very close to belay points. Remember that, although speleo ropes are made to resist wear, *all* ropes will abrade when rubbed against rock, and so *must* be protected. Tackle bags are the best way to transport ropes and these can often be put to good use as rope pads when empty. Purpose-made rope pads are usually a long strip of heavy-duty material which can be closed around the rope with 'velcro' strip fastening.

Much thought and advice has been expended on the question of when to 'retire' ropes. If they have been kept out of direct sunlight there is no need to discard them simply because they are two or three years old. The degree of visible wear is a better indicator. If a rope is all 'furred out', chuck it away. Nylon tends to harden with age, but with great wear the rope may start to slacken again; this is the time to throw it away. No hard and fast rule can be given, but 'If in doubt, throw it out' is good advice. Remember: it is your life you are hanging on the line!

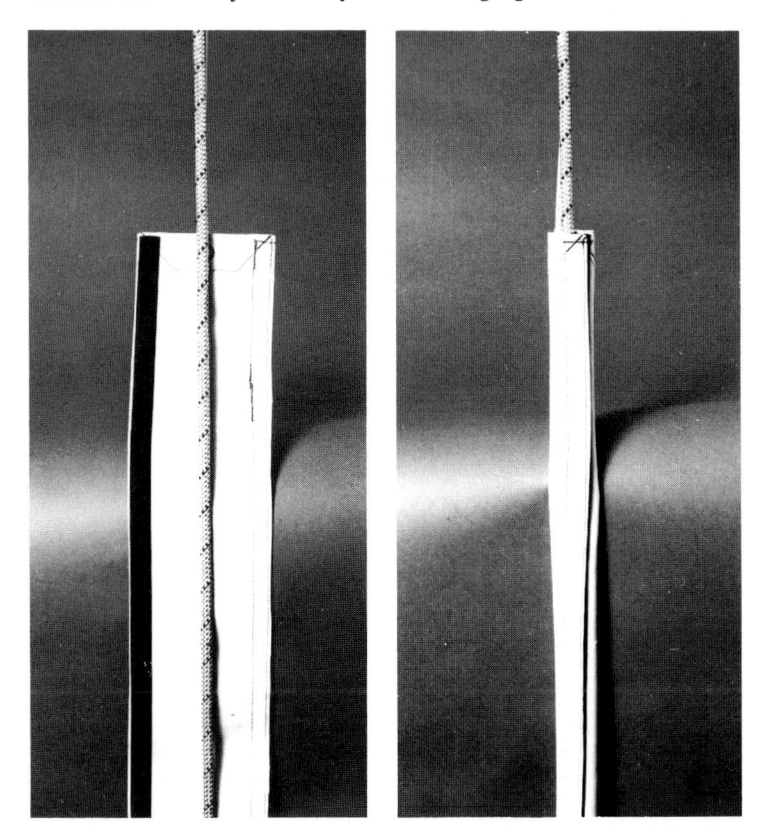

Rope-pad. The rectangular pad, of heavy, tough material, is closed with a strip of velcro, and attached to the rope with a small prusik loop

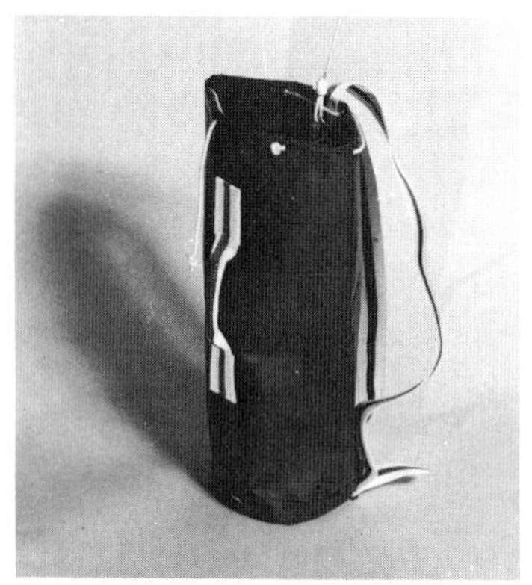

Tackle-bag. Can be carried as rucksack or kit-bag, with reinforcing round the wear-points

Packing rope in a bag. First tie the rope end to the bag, then:

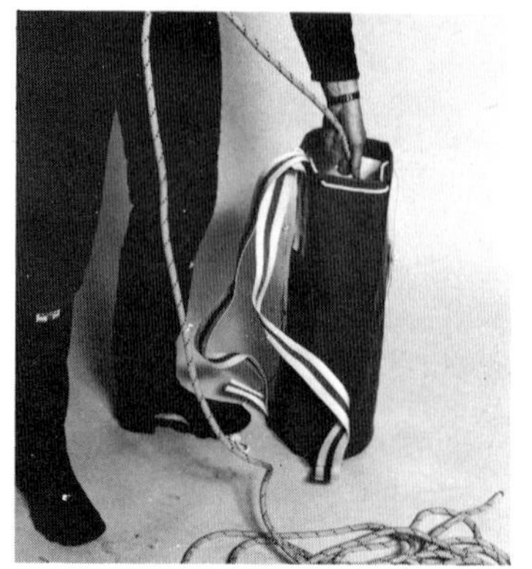

Feed the rope loosely into the bag, shaking it down from time to time

Rope coiling. Rope loosely piled on floor, end held in left hand, loop gauged by arms at 45°

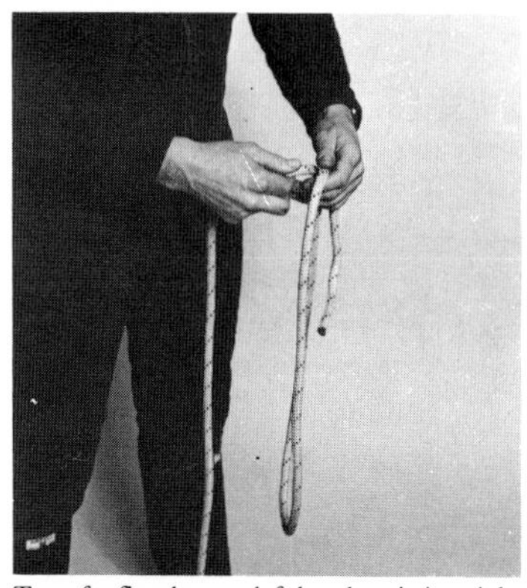

Transfer first loop to left hand, twisting right hand to keep coil flat

When less than 2m of loose rope is left, take the short end and loop it back as shown

Take the long end and wrap it around the coils, trapping it on itself

Thread the second end through the loop in the first

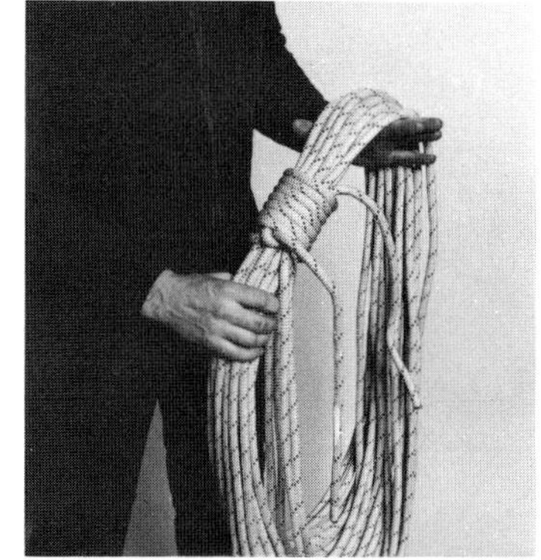

Tighten everything up

Rope bags (or tackle bags) have already been mentioned. These vary from simple drag-bags to deluxe underground rucksacks. The former suffice for short systems, but the latter come into their own on very long underground expeditions, as they can be either dragged or carried on your back, and are less likely to fall to bits at critical moments. The most useful size carries 100m of rope, weighing around 10kg when full and wet. At my own caving centre we coil ropes for storage, transferring them to bags before a 'trip'.

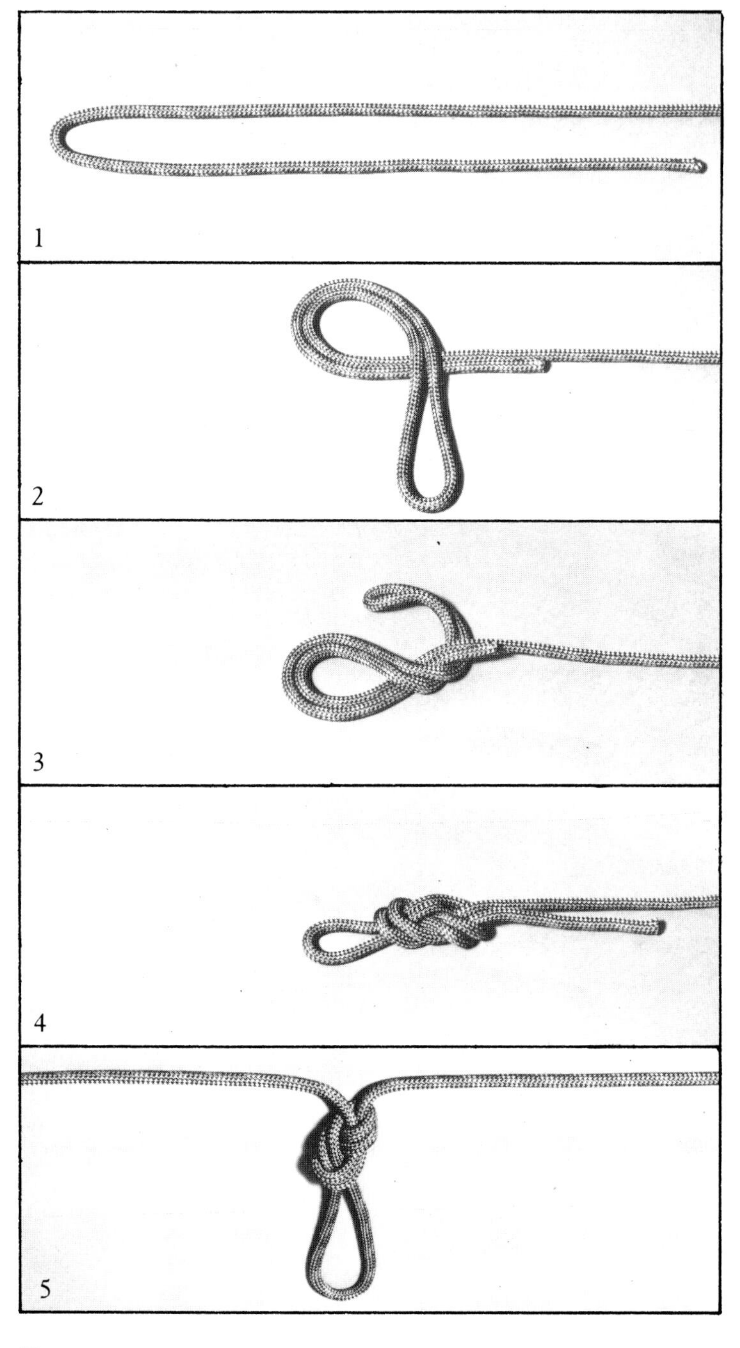

1-4. Double figure-of-eight knot.
5. The knot can be tied in the middle of a rope as well as the end

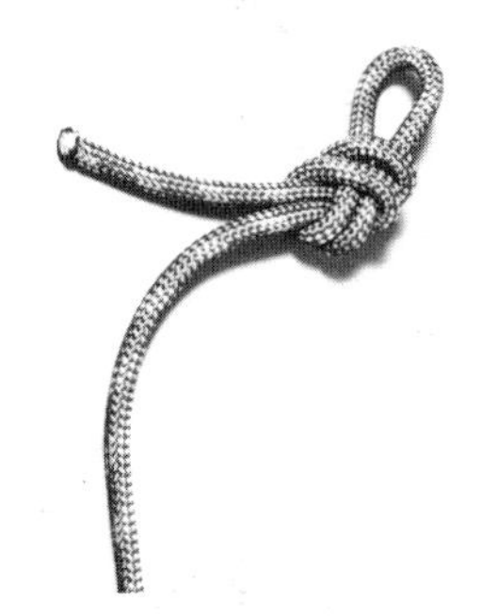

Double overhand loop. One less turn than the figure-of-eight loop. Can come undone easily, or, conversely, bind tight

Join two ropes by tying a figure-of-eight knot loosely in one rope end, and 'chase' the other rope through it. Leave 1–2 metres spare in the upper rope end, so that a loop can be tied in it to clip into if the rope join has to be passed in mid-pitch

Knots

Before using ropes a basic knowledge of knots is essential. Here are some you'll need to know.

Double figure-of-eight knot: The most important knot of all, this provides a loop, made to whatever size is necessary, at the end or in the middle of the rope. The loop so formed is normally used to clip into a karabiner or Maillon Rapide attached to yourself or to a belay.

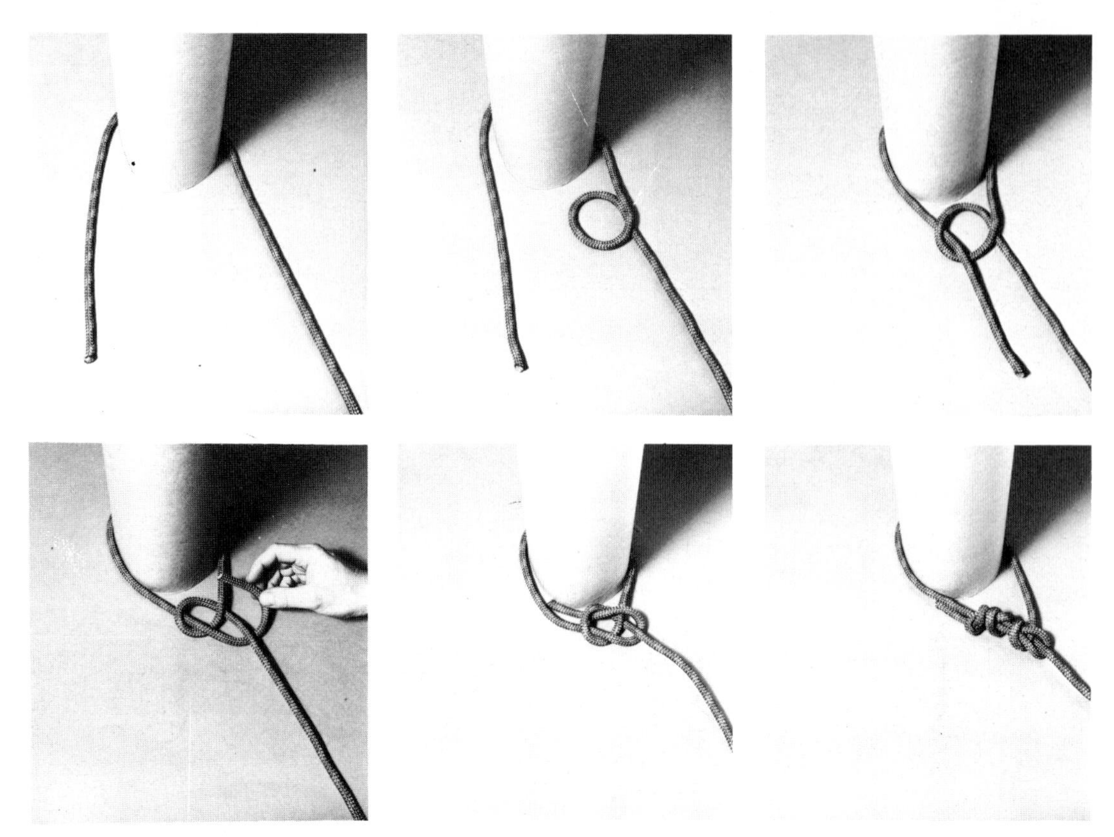

1-6. The bowline. For tying a loop around a bollard. Tends not to tighten, hence 'stopper knots' are essential (photo 6)

Bowline: This is another loop-knot, but can be tied only in the end of the rope. It is more convenient when the end of the rope has to be passed round a 'thread' before the knot is tied, and is easier to adjust. It can work loose, so the 'stopper knot' is essential.

Tape knots: Slings made from nylon tape are a very useful accessory, particularly when natural belays are used. They can be purchased made up, with a sewn join, or made at home by knotting the two ends. This *must* be done with the tape knot — other knots are dangerous.

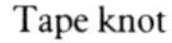

Tape knot

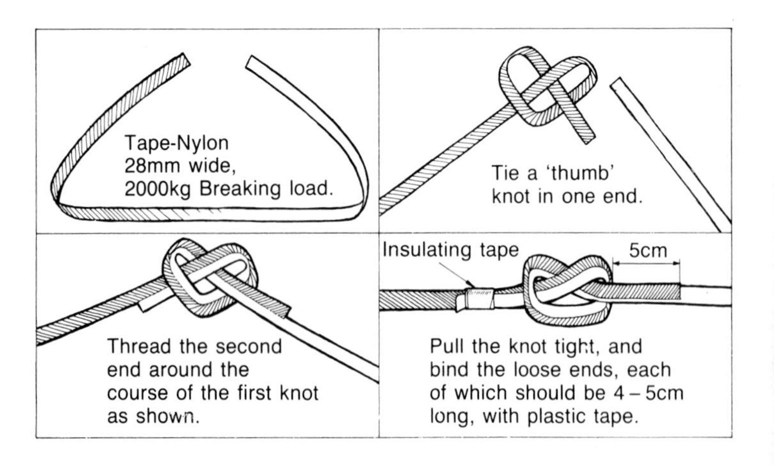
Tape-Nylon
28mm wide,
2000kg Breaking load.
Tie a 'thumb'
knot in one end.
Thread the second
end around the
course of the first knot
as shown.
Insulating tape
5cm
Pull the knot tight, and
bind the loose ends, each
of which should be 4 – 5cm
long, with plastic tape.

illustrated p.87
Little Hull Pot, Northern Pennines, UK

Harness

Leaving knots aside for a while, let us turn our attention to the equipment which the caver wears, and which allows progression up and down the rope.

Simplest of all is the load-bearing belt. Made of 50mm tape, with a locking buckle, such belts can carry the accumulator of a miner's lamp (see pages 65–66), and can be used to clip the wearer onto a rope. They are widely used in climbing ladders, when you have a lifeline clipped to a load-bearing belt by a karabiner. However, if there is any chance of your whole weight having to be suspended from a belt there is a danger of internal injury; and, even for lifelining use, my advice would be to add a sit-harness to the belt, so that in the event of suspension you will be sitting comfortably, supported by straps under the top of both legs. The sit-harness is essential as soon as you progress to climbing and descending ropes. There are various types:

Leg-loops: These are used in conjunction with a load-bearing belt. Although not quite as comfortable as a complete sit-harness, they have the advantage that you need no extra belt to carry your accumulator. To secure leg-loops and belt together and form the main body attachment point, a 10mm delta-shaped Maillon Rapide is used. This is like a triangular chain-link, one side of which has a screw-sleeve which opens and closes. When screwed shut, Maillons can be subjected to load in any direction without loss of strength, so they are much safer for this purpose than karabiners. (They come in steel and alloy — either can be used.) A disadvantage of a simple leg-loops-and-belt outfit is that the Maillon Rapide sticks out at right angles to the body: this can be a snag in tight caves, and leads to problems of alignment of the chest jammer (see below), which is clipped into the Maillon.

Speleo sit-harnesses: These overcome the alignment problem. The three types illustrated give a good idea of the range available.

Leg loops and belt. Front and rear

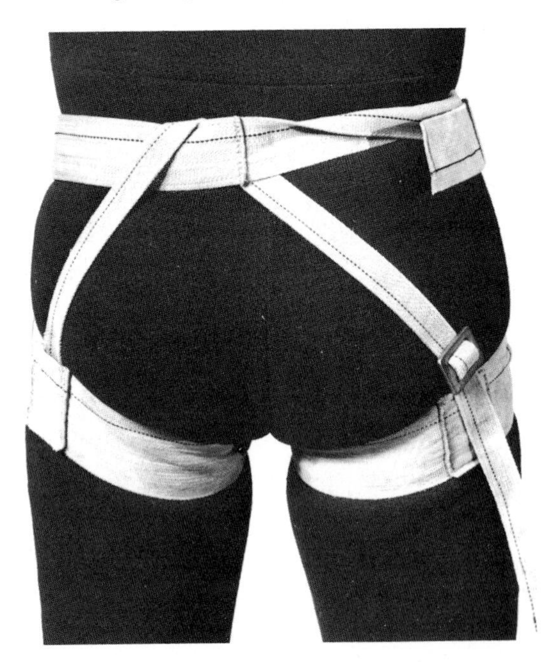

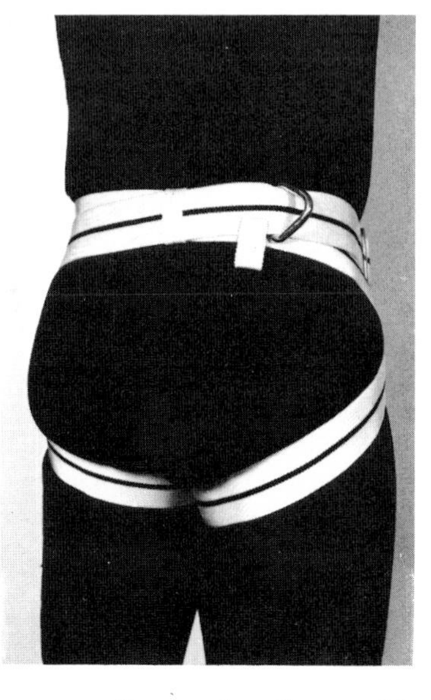

Classic type sit-harness, front and rear

Classic harness: This gives great comfort and safety, with completely adjustable leg-loops attached to two independent body straps. The outfit is ideal for long pitches, but slightly cumbersome for wearing around a lot of caves, and fiddly to re-adjust.

'Rapide' harness: This consists of a belt, with a lower (bum) strap which is held in place by a tape which threads through it from the front. Although not quite as comfortable as the 'classic', this harness can easily be relaxed for comfortable movement between pitches. I have used one in the very hot caves of Sarawak, and found it ideal for the occasional pitch.

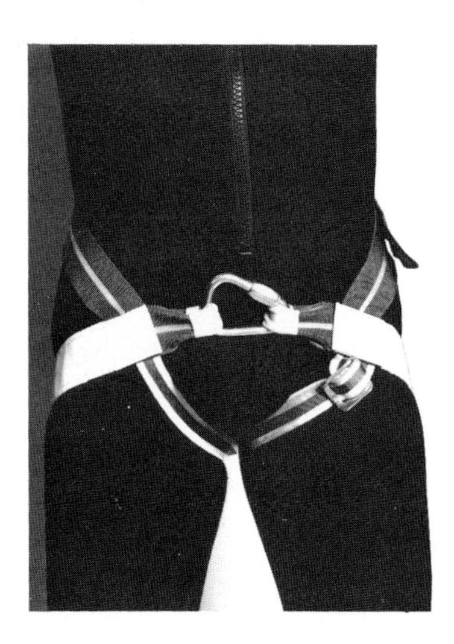

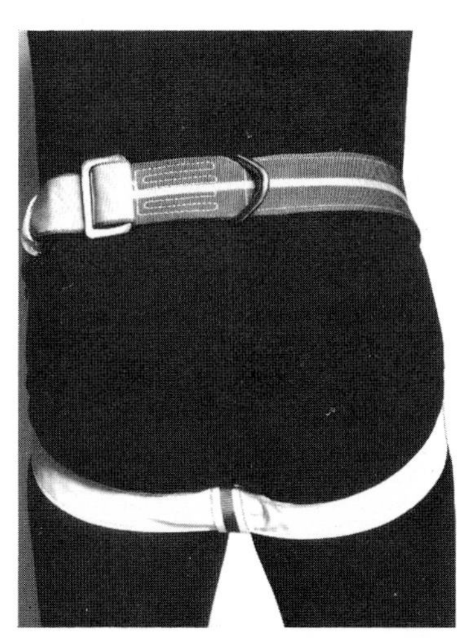

'Bum-strap' type harness. The sit-strap is held in place by a secondary strap which can be unbuckled for ease of travel in horizontal passages

'Avanti' harness: Recent trends have been towards lighter weight gear, and this is the answer for those wanting a lightweight harness. Made of 28mm tape, it can be adjusted in seconds. It holds the 'chest' jammer in a lower position than any other harness, a technical point best appreciated by the experienced! This outfit is the best buy for the expert.

As with ropes, looking after harnesses is just as important as selecting the right one. Beware of cheap harnesses: the quality of tape varies greatly from harness to harness, and the best types, which do not lose all their strength if an edge or the surface is abraded, are expensive. Make sure that the obvious wear-points are protected. Keep your harness washed clean, and discard it as soon as you have the slightest worry about wear. Remember that tapes lose strength more rapidly with wear than ropes, and that badly furred or cut tape is potentially dangerous.

'Avanti' lightweight harness

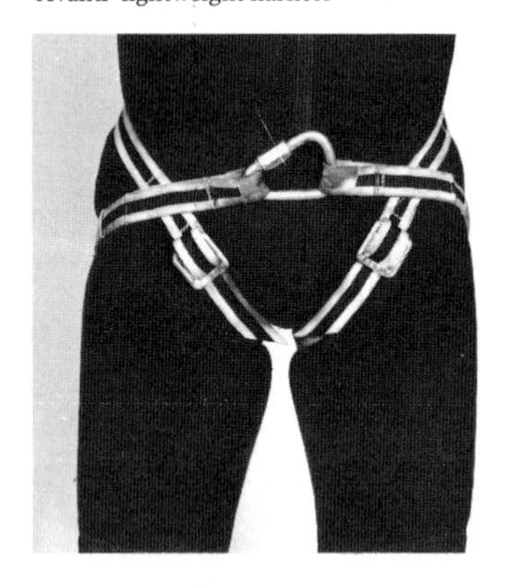

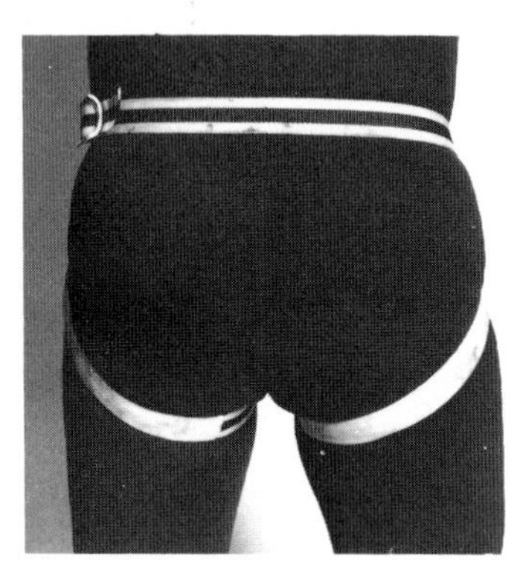

The sit-harness and Maillon Rapide are all that is required to clip a descender in for a descent, but for coming up again a chest-harness is needed as well. This is *not* a major load-bearing component: if a major part of your weight is supported at your chest, your system is dangerous. Therefore I recommend a simple chest-strap only, the figure-of-eight chest-strap, which has a 2.75m length of 25mm tape with a one-way sliding buckle. Fixed as shown in the diagram, it can easily be tightened when you are actually climbing pitches, and loosened for movement in between.

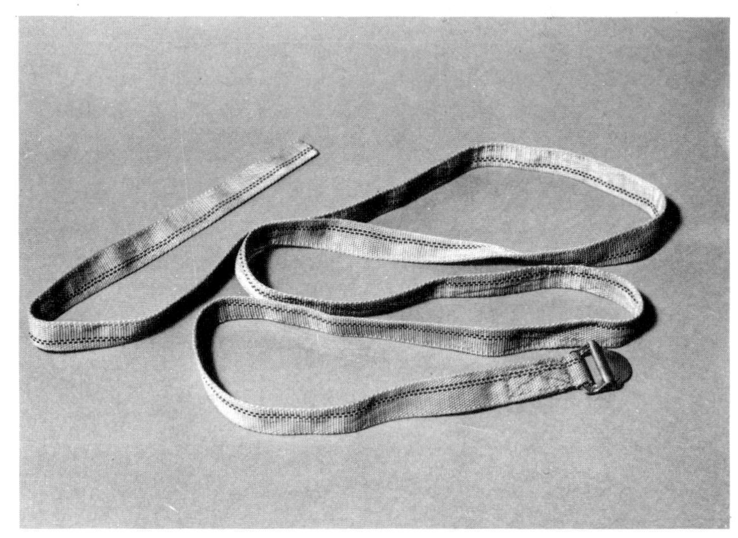

Chest strap. Simply 3 metres of tape with a quick-adjust buckle

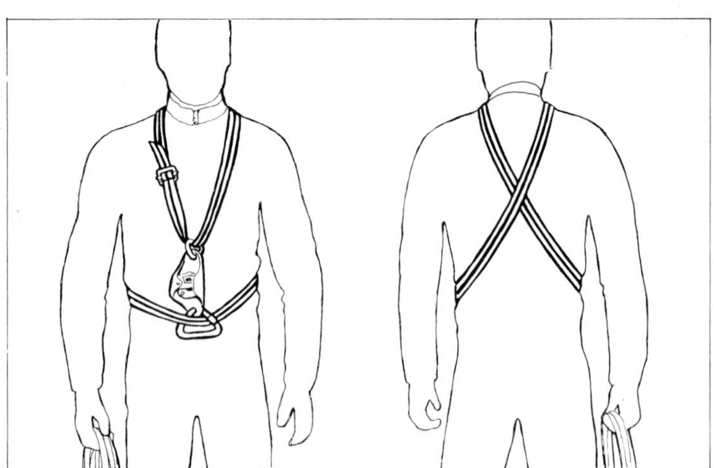

Chest strap in position

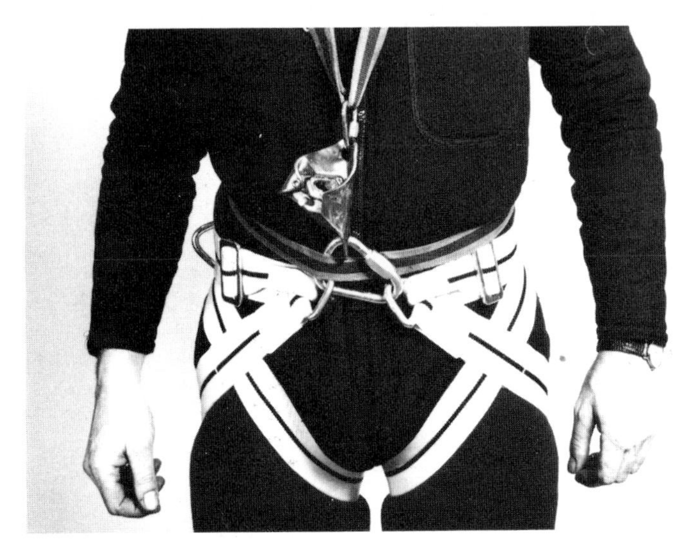

The full harness in place, with the chest (croll) ascender

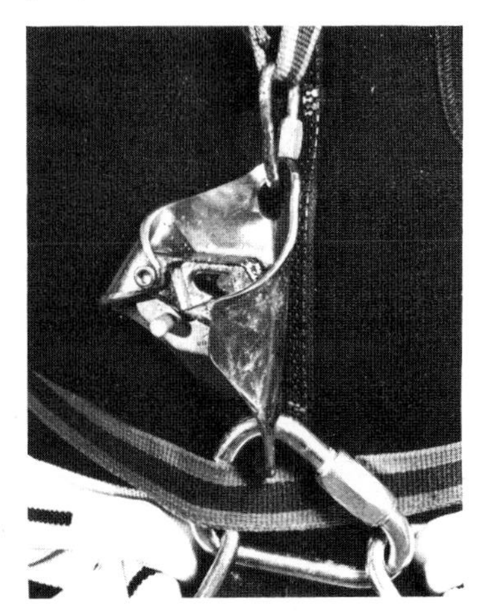

Close-up of croll, clipped into the main attachment (10mm Delta Maillon Rapide)

Jammers

Jammers are metal clamps that are used for climbing up ropes. They slide freely if pulled up a rope, but fasten firmly onto it if pulled downwards. They are always used in conjunction with a harness arrangement. Reference has already been made to the chest-jammer, which is fixed between sit-harness and chest-strap. You must be able to get the rope in and out of it without having to undo anything.

The Petzl 'croll', made for this purpose, is widely used. It has a large hole at its base which allows free movement of the Maillon Rapide, and sloping teeth which minimise friction as the jammer is pulled up the rope; it is engineered so that it lies flat against the chest. Other jammers which can be used in this position are the Petzl handled jammer, the CMI and the Jumar.

With one jammer positioned above the sit-harness, you can sit comfortably halfway up a rope, but something more is needed for progression. The two main techniques currently in use are the sit-stand (or 'frog') and ropewalking. Each of these has its variants.

Alternative jammers which can be used for chest-mounting:

1 'Clog' expedition jammer

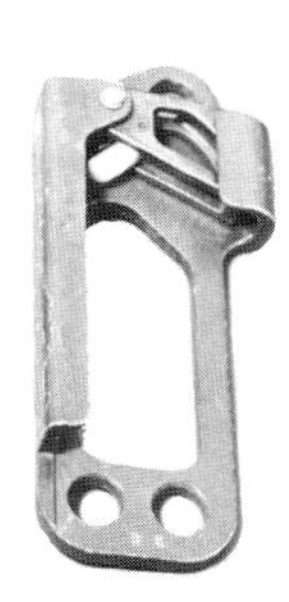

2 CMI jammer

3 Petzl handled jammer

4 Jumar

For the 'frog' — which I recommend as the system to learn first, and continue to use unless you are regularly doing 100m-plus pitches — you need a second, hand-held jammer. Again it should be possible to clip it to and unclip it from the rope while it is still secured to you. This can be done with the Petzl jammer and the Petzl handled jammer, the CMI and the Jumar. I use a simple Petzl jammer; having an actual handle tempts you to clutch the device and rely on arm strength rather than your legs.

Ropewalking requires jammers which run completely free of the rope when pulled up, and for this purpose a US design — Gibbs, or one of its derivatives — is universally used.

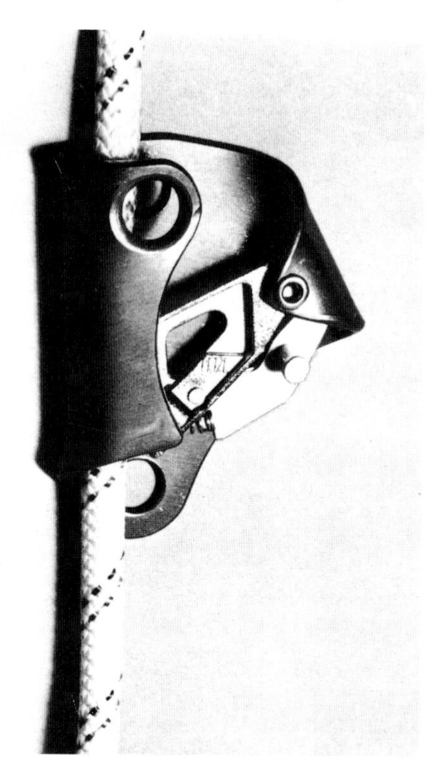

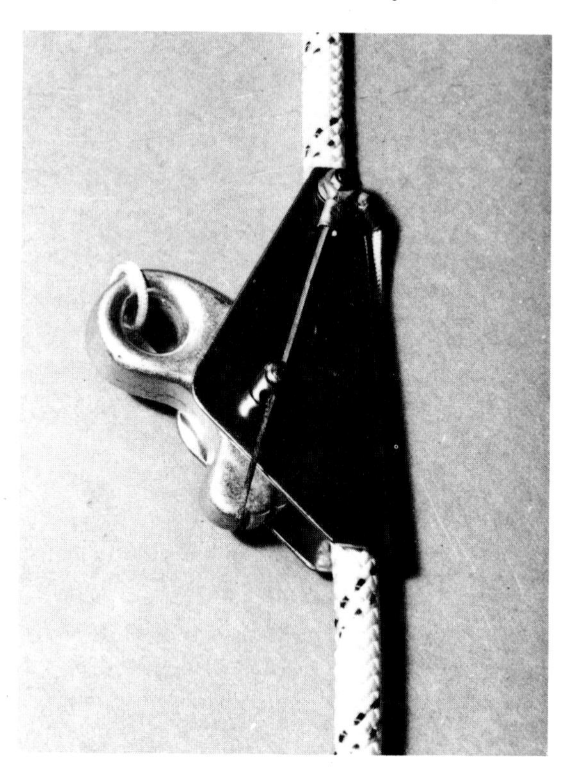

The Petzl jammer is simplest for use with a foot loop, though all the handled jammers can also be used

Ropewalker. The fastening point is the end of the cam, rather than the sheath, as on all the other jammers illustrated

Descenders

The last major item in a personal 'rig' is the descender, used for sliding down fixed ropes — preferably under control! Types suitable for caving should be loadable while clipped to the caver, should not put a cumulative twist into the rope, and should not put the caver in danger if wrongly loaded.

Continental cavers favour the Petzl bobbin descender: the normal version requires skill to 'lock off' in mid-descent, whereas the 'stop' version locks itself. Both are light and smooth in use. Cavers in the US use the Rack. This comes in a number of variations, but all work by bending the rope around alternate sides of metal bars threaded onto a long U-shaped rack. Safe use requires more skill than with the bobbin, and no self-lock model is available.

There are other descenders available, but to try to deal with the whole range would only confuse the issue. (See illustrations.)

Petzl bobbin descender

In normal use the rope is threaded from the anchor point around the bottom bobbin from left to right, and then round the top bobbin. From here the rope is clipped into a second (steel) karabiner, which gives extra friction near the bottom of pitches. The right hand controls descent

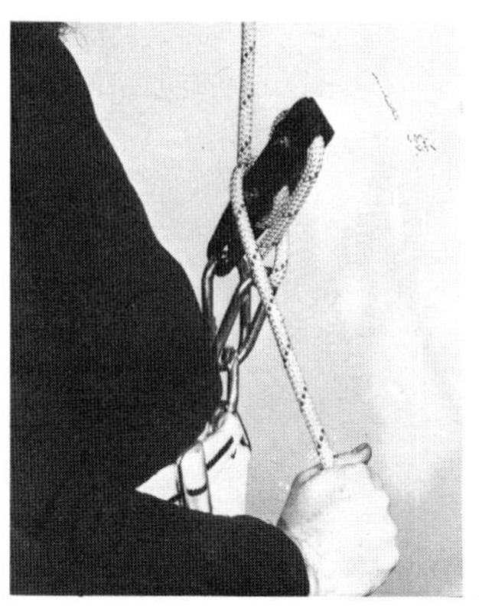

To lock the descender off the tail-rope is notched between the loaded rope and the descender. (Soft-lock — still needs holding.)

Before the control hand can be taken off the rope (hard-lock) the tail-rope must be looped through both karabiners, and the loop hooked over the descender

5-bar rack

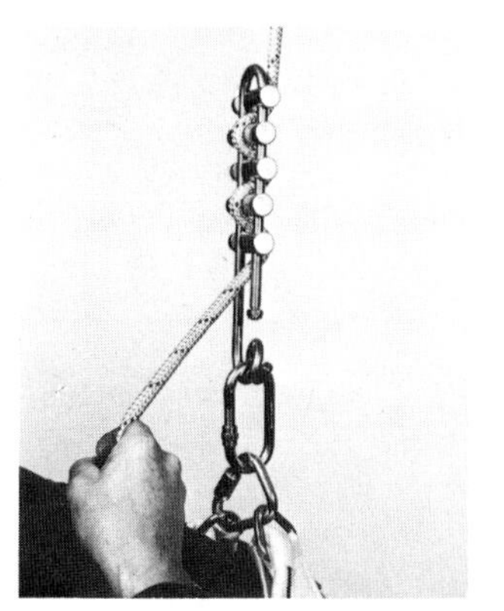

The rope is threaded as shown. Control is by holding the rope below the rack, *and* sliding the bars up the rack to trap the rope

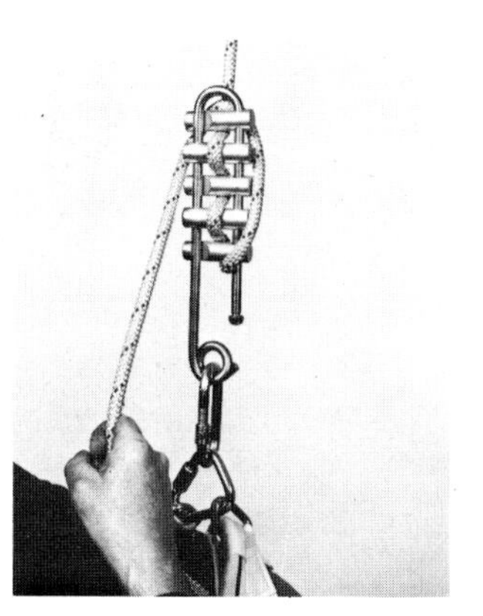

By pulling the rope up and notching it over the rack-top a 'soft-lock' is achieved

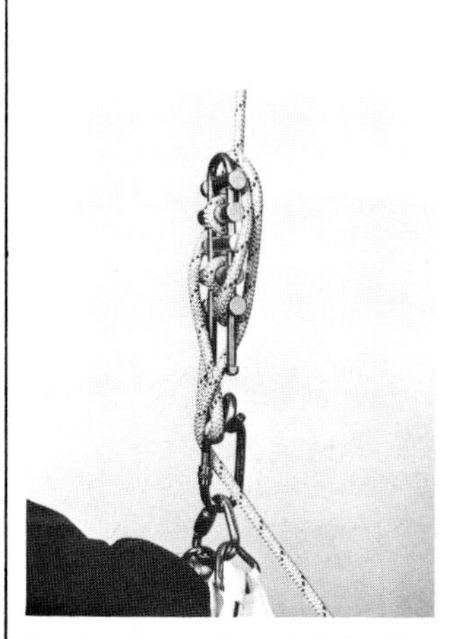

Looping the rope through the karabiner and over the rack-top completes the 'hard-lock'

'Stop' self-lock descender

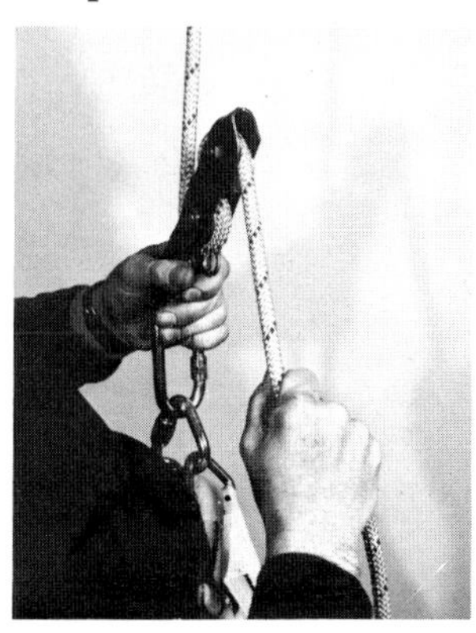

To move down the rope, grasp the handle and body of the descender together, which removes the lock, holding the rope below the 'stop' in the right hand, and controlling descent as with the Petzl bobbin

The 'Stop' exposed. When loaded, the bottom capstan pinches the rope against the top one and must be held open with the handle

With 'hands off' the 'stop' locks itself

Bits and pieces

There are still some important items to be added to your equipment before you are all girded up.

'Cow's tails'

This is the name given to the security strops which are fastened into the main attachment maillon. To make the normal double model, take 2.5m of 11mm *climbing*-rope (for extra shock-load absorbency). Tie a small double figure-of-eight loop in each end,

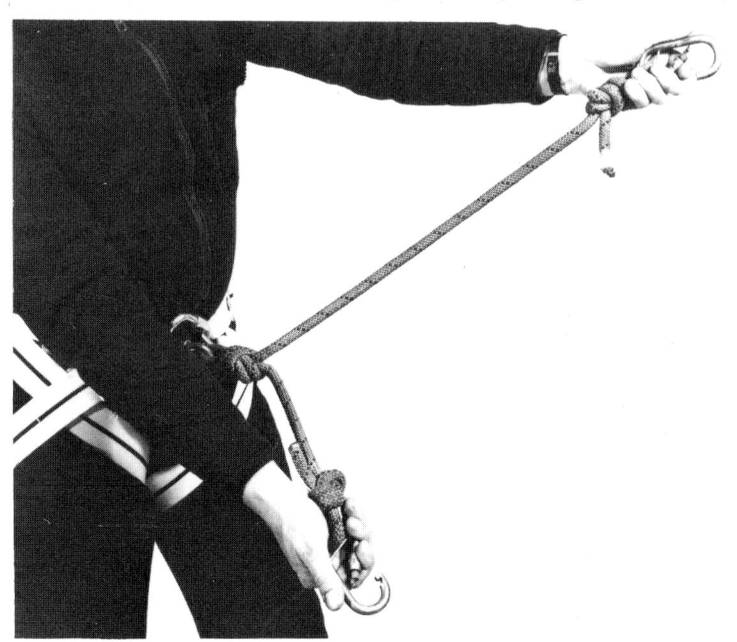

Set of long and short cow's tails, with karabiners

Karabiners

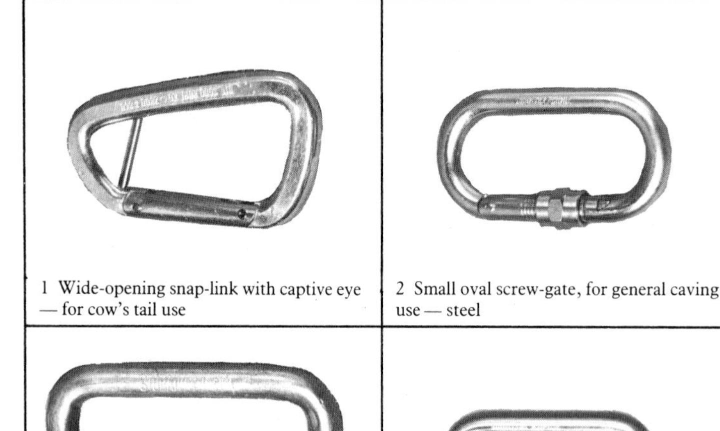

1 Wide-opening snap-link with captive eye — for cow's tail use

2 Small oval screw-gate, for general caving use — steel

3 Small oval screw-gate, for general caving use — alloy

4 Large offset screw-gate. One clipped to the waist can be useful

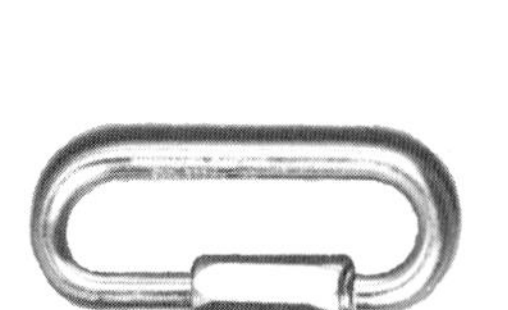

5 7mm long-opening maillon rapide. A light, cheap substitute for karabiners for many purposes

6 Large pear-shape — for Italian Hitch belaying

and c [illegible] wn in the
illust[illegible] ey have a
thin [illegible] the same
effect [illegible] pe. They
have a [illegible] sy to clip
into r[illegible] a double
overha[illegible] nish with
two st[illegible] he other
about [illegible] directly
to the [illegible] ness. (A
smaller [illegible] eded for
this.)

The foot-loop length should allow the jammer attached to it to be positioned just above the chest jammer when standing upright

The security link length MUST allow the jammer to be reached when it is at full stretch

Foot loop

A ***foot loop*** is needed to make use of the jammer which is not clipped directly to the body (see page 91). It can be fairly thin low-stretch rope — 8mm low-stretch terylene is suitable. Some cavers like a slightly more complicated model with a loop for each foot, but my own preference is for the one-foot model, made with a loop large enough to take both feet if desired. The length of the loop attachment point is very important. When standing upright with your foot in the loop, your hand-held jammer should be just above the level of your chest, or body, jammer. This allows the maximum 'gain' with each step upwards.

When not in use one's SRT kit can be carried in a small waist-bag

A second double-ended strop, this time made of 9mm climbing-rope, links hand-jammer to sit-harness. When in place, this ***security link*** should allow you almost — *but not quite* — to stretch your arm upwards to its limit. It should *always* be used when ascending.

Karabiners are widely used for clipping components together. Most useful to the caver are small oval ones with a screw-gate. The small ones are sufficiently strong, and their oval shape allows equipment to hang centrally from them. Alloy karabiners can be used for all purposes except being clipped into a 'running' rope, when abrasion can saw through them. Steel ones, which are slightly heavier, can be used for everything.

A final item of personal gear is the ***kit-bag*** into which you put everything else. A simple model which can be clipped to the waist and closed with a draw-string suffices.

Fixed rope techniques

Anchoring

Having loaded our rope into its bag and kitted ourselves out, one thing remains before we can start our descent — belays for rigging the pitches. Whenever possible, natural belay points should be used — in conjunction with rope-pads (see page 83) where necessary. However, probably more often than not nature fails to provide secure threads or bollards in the right places, so artificial anchor points have to be engineered to avoid rope abrasion.

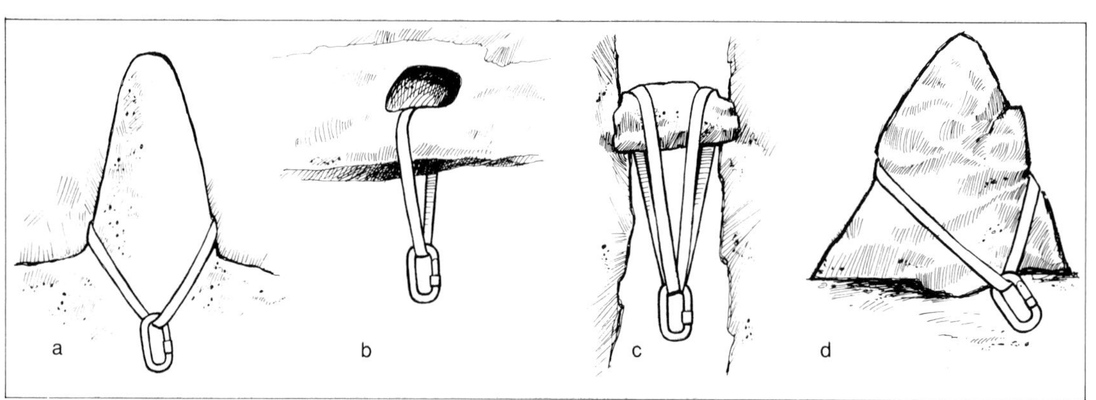

Natural belays

a. Stalagmite. (Should have minimum diameter of 80mm, be used at the base only, and be founded on rock, *not* mud!)

b. Natural 'thread'. Corroded limestone often contains usable holes

c. Chockstone. Ensure that the direction in which a jammed stone will be pulled will wedge it further!

d. Boulder. Make sure it is big (at least 500kg) and well founded!

Cave conservation is important in every country, and attitudes differ as to the propriety of such aids. My own view is that the best way to damage any cave is to have an accident down it. Immediately someone is injured, priority is given to getting the victim out, and if the cave is damaged in the saving of human life then so be it. Inserting safe anchor points does minimal damage to the environment, and may prevent greater damage being caused in the aftermath of an accident. However, it is important for even inexperienced cavers to be able to distinguish a good anchor point

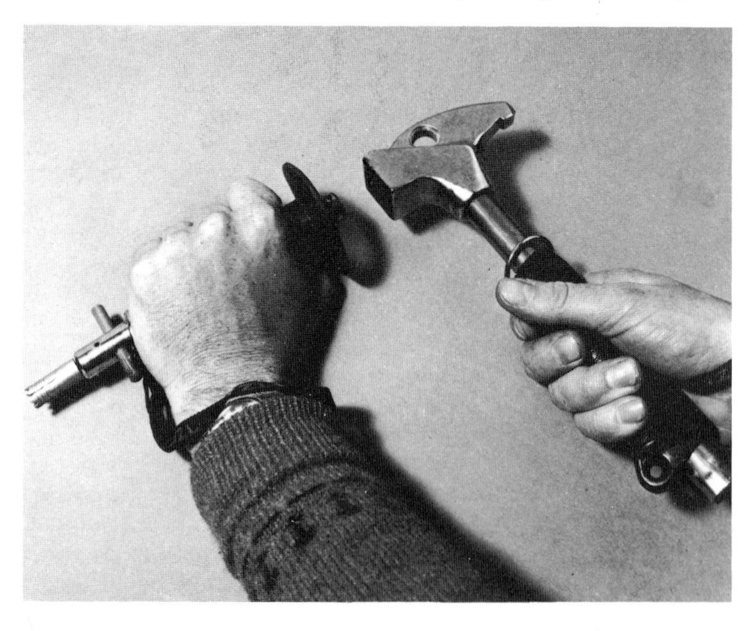

8mm drill anchor, screwed onto driver ready for drilling

from a bad one if the peppering of pitch-heads with superfluous bolts is to be avoided.

The first (back-up) belay should be well back from the pitch head, and if possible rather higher than the next (hang) belay. A natural back-up belay can usually be found. The hang belay, on the other hand, should allow the rope to hang free of any obstruction, and for this reason an artificial bolt will usually be required. The most common size and type in current use is the 8mm self-drill anchor (see illustration on page 96).

Belay positioning

a. Back up belay. Must be bomb-proof. Often a natural thread (preferred) or spike

b. Hang belay. Must be positioned to allow a free hanging rope. Usually an artificial anchor and bolt

NB. Never place complete trust in one 8mm bolt. Double them up as necessary

c. Two bolts used as back-up belay

d. Two bolts used as hang belay

The anchor is screwed onto the driver, and the best place for the anchor selected. The need to allow the rope to hang free has been mentioned, but the anchor must be placed in sound rock, with no cracks nearby, and away from any edges. Drilling into the rock is best done with light, tapping blows, rotating the driver at each blow (heavy hammering simply breaks off the drilling points), ensuring that the hole is at right-angles to the rock surface. The driver should be pulled out of the hole frequently so that the rock-dust can be tapped from its centre (a thin plastic tube can be used to blow rock-dust out of the hole if necessary). Drilling continues until the hole can be made no deeper: *never* trust a half-drilled hole. Now the metal wedge is placed in the end of the drill-anchor, and the assembly tapped firmly into the hole: this causes the drilling end of the anchor to splay out and grip the walls of the hole. Remember that if sledgehammer blows are used, the splaying can go too far and shatter the rock. If the anchor has been placed correctly, you will now have a well-placed threaded metal tube at right-angles to, and flush with, the rock face.

Drilling a hole to secure a re-belay point

An 8mm drill-anchor in place. It should have no cracks within 15cm, and not protrude from the rock

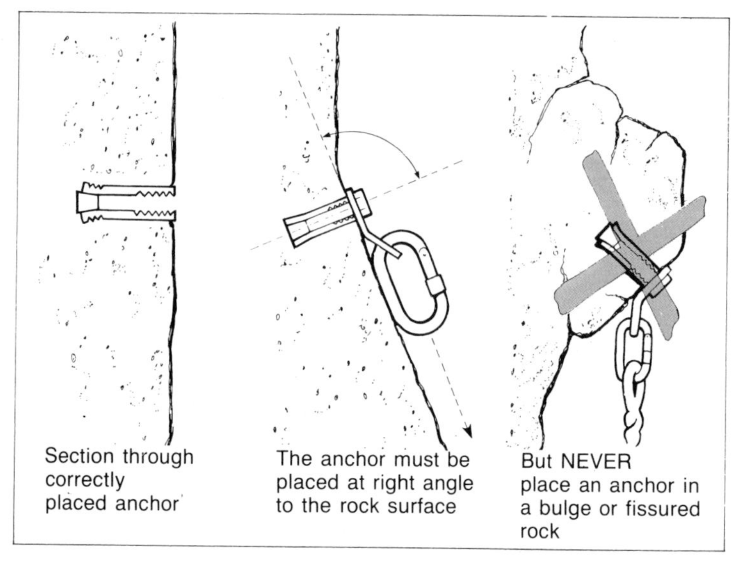

Section through correctly placed anchor
The anchor must be placed at right angle to the rock surface
But NEVER place an anchor in a bulge or fissured rock

Anchor placement

In newly discovered caves you will have to place your own anchors, but in many countries, particularly in Europe, known systems will probably have anchors in position already. If these are sound they should be used. Knowing just how anchors should be placed will help you to decide if existing ones are likely to be 'good'.

The anchor is used by screwing a 'hanger' into it. These come in a baffling variety. I will describe just three types, which between them cover all needs.

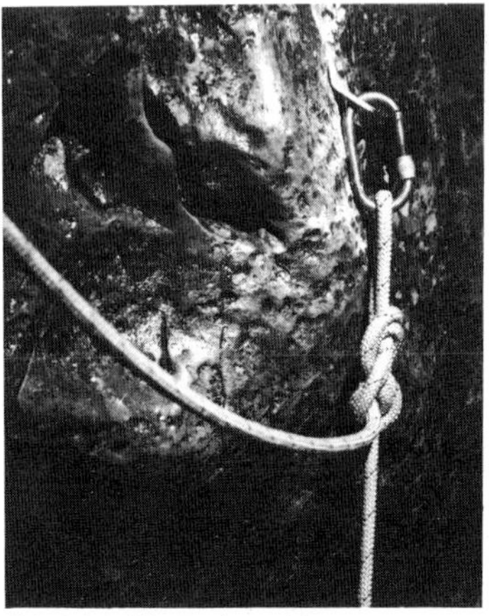

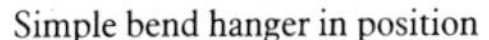

Simple bend hanger in position

Half-twist hanger

Ring hanger. Note the rope looped directly into the hanger

The ***half-twist hanger*** is a metal plate with two holes. The upper one holds the bolt; the lower one, positioned at right-angles to the upper, takes a karabiner or a Maillon Rapide, into which the rope is clipped. The geometry of the situation is such that the rope loop clipped into the karabiner is at right-angles to the rock face. If the face is overhanging slightly, this rope loop will hang clear of it, and all is well; but if the face is just vertical, the rope might rub, and a ***simple-bend hanger*** is preferable. If the anchor has to be drilled upwards into a roof, a ***ring hanger*** should be used, as the other two types would tend to lever themselves out. However, the ring hanger is weaker, and must be used with care. The first two types of hanger require a Maillon Rapide or a karabiner to link rope and hanger, while the rope can be linked directly into the ring hanger.

When screwing hangers into anchors, never over-tighten the bolt as this could strip the threads, with potentially disastrous consequences. Check pre-existing anchors carefully before you use them to make sure that their threads are sound.

Safety at the pitch head

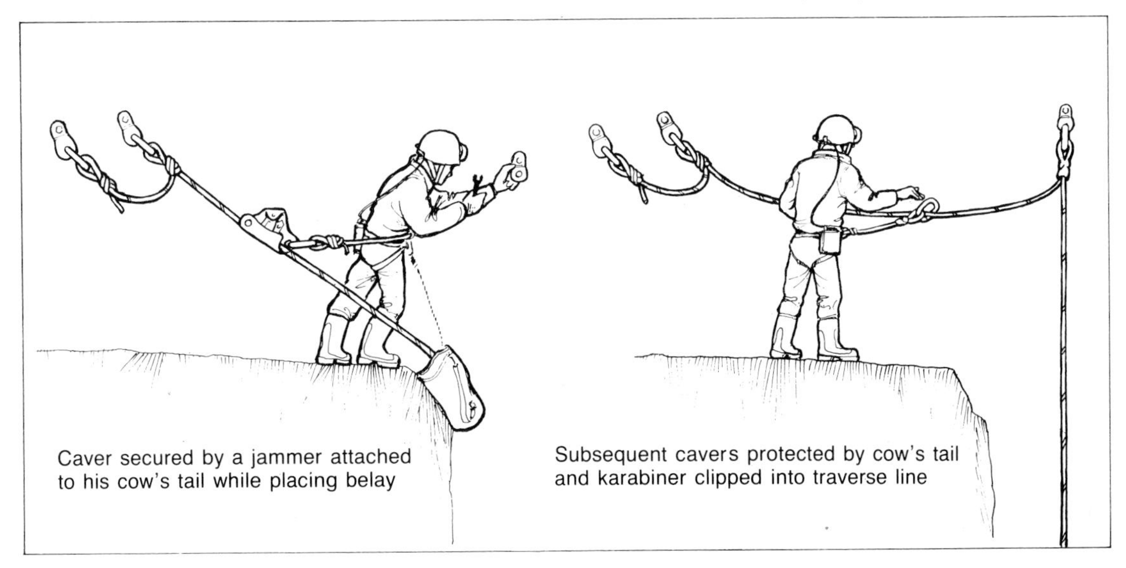

Caver secured by a jammer attached to his cow's tail while placing belay
Subsequent cavers protected by cow's tail and karabiner clipped into traverse line

Descent

Enough about gear. We can now proceed into our pothole system and start to rig the first pitch.

Before reaching the lip of the pitch the leader selects a good belay. In some countries this will usually be a bolt; in the UK there are usually good natural spikes or 'threads' to hand. The top end of the rope is taken from the bag and secured around the belay, or to a tape sling fixed to the belay. The pitch-rigger clips himself to the first belay with a cow's tail, or, if the hang belay will be some distance away, clips a jammer onto the rope and clips his cow's tail into that.

Now the second belay, the main hang belay, is selected and placed. As indicated earlier, if this is to be a free-hang it will usually be a bolt; if not, padding of the pitch lip is vital. It may be possible to get the rope to hang 'free' for only part of the way down the pitch, and in such a case the rope will have to be re-belayed at the potential abrasion point. For the time being we shall consider the procedures to be adopted on a straightforward pitch.

The first caver clips the rope bag to his main harness attachment position on a cord long enough to let it dangle just below his feet. He now attaches his descender, already clipped into his harness, to the rope, taking great care to load it correctly. It should then be locked off. (See diagrams.) The caver can now sit with his weight on the rope and check finally that everything is as it should be *before* unclipping his cow's tail. Next, with great care, he unlocks the descender, taking particular care to ensure that a good hold is kept on the end of the rope *below* the descender. If a rack is being used, at least five bars should be loaded at the outset. On a long pitch, with a heavy weight of rope below, it may then be necessary to take the rope off one bar and ease the remaining bars down the rack frame before you can descend. Be very careful, though. As you descend a pitch, and the weight of rope below the descender decreases, so does the friction

Rope pad in place at belay. It is acceptable to pad a rope *immediately* beneath a belay, but not anywhere else, as it may slip off when the rope moves

Starting the descent. When starting an abseil, check that everything is OK before unclipping the cow's tail and starting descent

Even experienced cavers are well advised to practise use of new items of gear above ground first. For beginners it is essential!

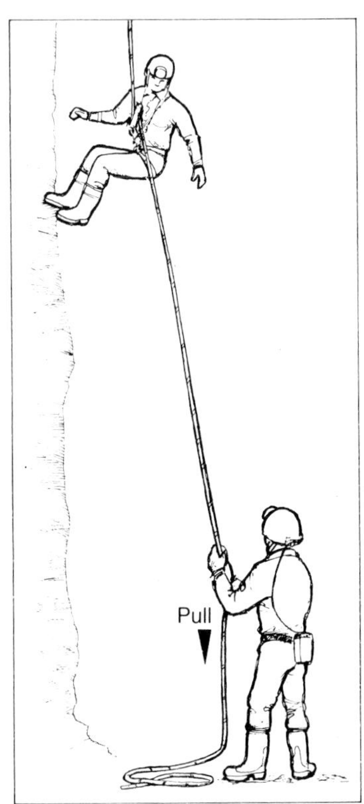

Bottom lining. A safeguard against loss of control by the abseiler. The caver at the bottom should not be directly underneath, and should hold the rope in both hands ready to pull if necessary

of the rope as it passes through the descender, and a firm hold and a slow speed are needed to retain control. (The first caver does not have this variation in friction as he descends with the rope uncoiling from the bag just below him.)

The ways in which different types of descender are controlled vary, and it is not possible in this short book to give such technical details. You should, before using any descender underground, have tried it on a rope hung from a tree or a roof beam, and thoroughly mastered its use a metre or two above ground level. You must be able to cope with a weighted rope (to simulate long pitches), to lock off the descender at any time, and be perfectly happy with the techniques. I would recommend that the beginner should first learn to use a self-locking descender — either the Petzl, as described, or, if it is likely that doubled rope will have to be used, the Lewis self-lock descender.

All self-lock descenders work on the principle that they grip the rope when the user lets go of the device. To descend, a lever is squeezed with the left hand, allowing normal descent control with

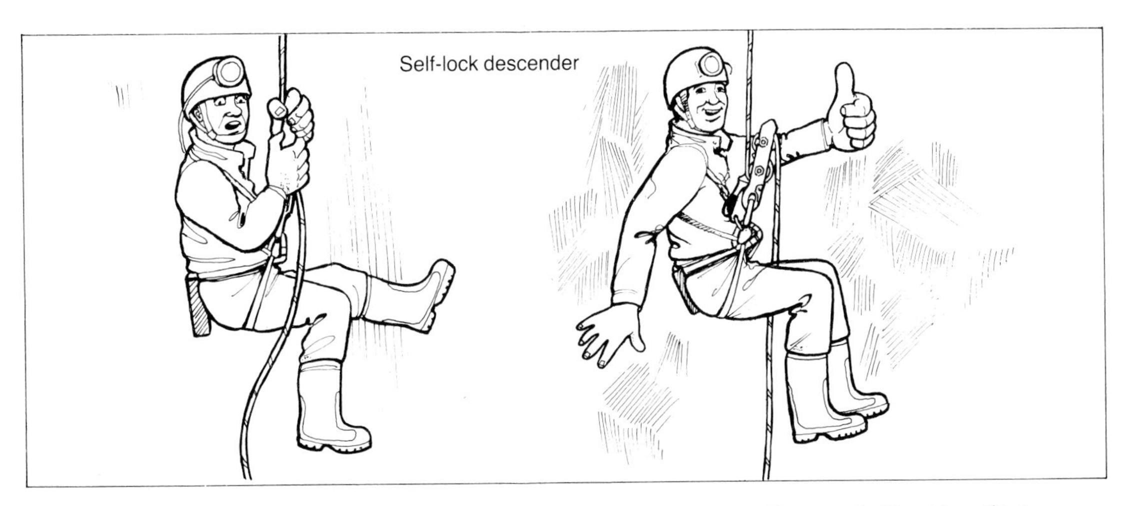

Emergency braking with a self-lock descender

the right. Uneducated panic can lead to a clutching of the lever, with the possibility of instant plummeting, so once again initial practice is necessary. Once the response *let-go-to-stop* has been mastered, self-lock descenders are both easier and safer to use.

When the first caver has reached the bottom, the second, who has approached the hanging rope with a cow's tail clipped into the loop between the two belays, makes his descent. It is possible for the caver at the bottom to safeguard the one on the rope: by pulling the rope-end, the friction on the rope running through the descender is increased, and so the sliding caver can be slowed or stopped.

Ascent

The principle of caving is usually the opposite to that of gravity: what goes down must come back up, unless there is a sneaky exit at the bottom of the cave. The process of climbing back up the rope is known as prusiking, after a certain Dr Prusik who invented the first practical technique for ascending. Dr Prusik used short loops of cord, about half the diameter of the hanging rope, as his 'jammers'. Each loop is threaded twice through itself in such a way as to trap the rope. By pulling on the end of the loop, the resultant knot is made to jam; but when the loop is slackened the knot can be pushed up the rope. For many years the Prusik knot and its derivatives were widely used by cavers, especially in the USA. Although largely superseded by mechanical jammers, prusik loops are still a valuable accessory, especially in emergencies, so it is worth practising their use and carrying a loop or two when on a vertical caving trip.

a. SIT — push foot jammer up rope

b. STAND

c. SIT — and repeat...

Sit-stand ascending requires a body or chest jammer held in place between sit-harness and chest strap. Properly adjusted, this allows you to sit comfortably suspended on the rope. Now the hand-held jammer is clipped onto the rope above the chest jammer, with the foot loop hanging from it and a security link between the sit-harness Maillon and the hand-held jammer.

The security link should *never* be overlooked. If one jammer should slip, come off the rope or otherwise fail, it is essential that

Prusik loop
a) Take a loop of cord, about half the diameter of the main rope

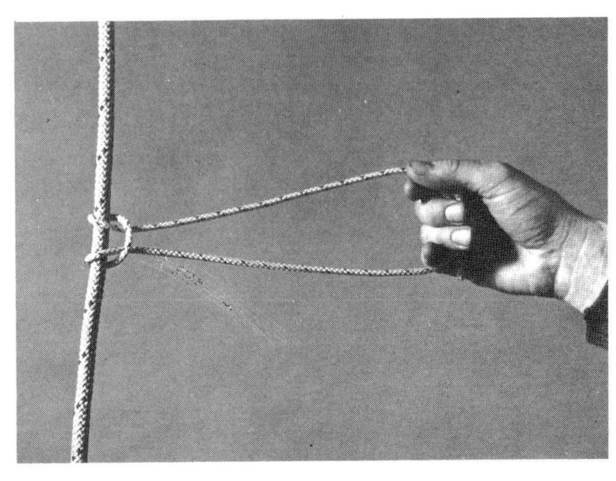

b) Loop it through itself, trapping the rope

c) Loop it through a second time

d) The resultant knot will slide if pushed directly, but locks when pulled by the loop

the other should support you comfortably, and this is exactly what the security strop ensures.

With all your gear in place, upward progression should be simple. Stand up in the foot-loop, pulling yourself in to the rope with your hands on the hand-held jammer. The chest jammer slides up the rope. You sit down and it jams, supporting you in your sit-harness while you slide the hand jammer up the rope and repeat the whole process. Nothing to it!

In practice, some find it easier than others. Try to keep your trunk in an upright position: to do this, it is important to use your arms to pull in to the rope, so that your feet push you up. A common mistake is to try to haul yourself up instead — very tiring and inefficient.

Sometimes, especially near the bottom of pitches, the rope can snag as you step up, so that instead of running through the chest jammer it just hooks up, and when you sit down you end up where you started. Trapping the rope between your feet while standing up prevents this, or you can simply ask the caver at the foot of the pitch to pull gently on the rope.

If the rope 'rides up' in the chest jammer instead of running through, try trapping it between the feet when standing

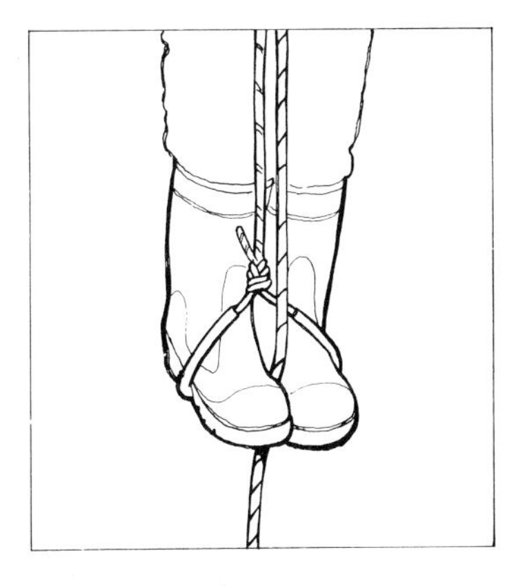

Progressing steadily, with rests as required, you should reach the top of the pitch without incident. Getting off safely at the top requires care. Standard procedure is:

- Clip long cow's tail into belay or loop of rope beyond.
- Stand in foot-loop and release chest jammer from rope.
- Step onto pitch head and release hand jammer.
- Move back to safe position before unclipping cow's tail.

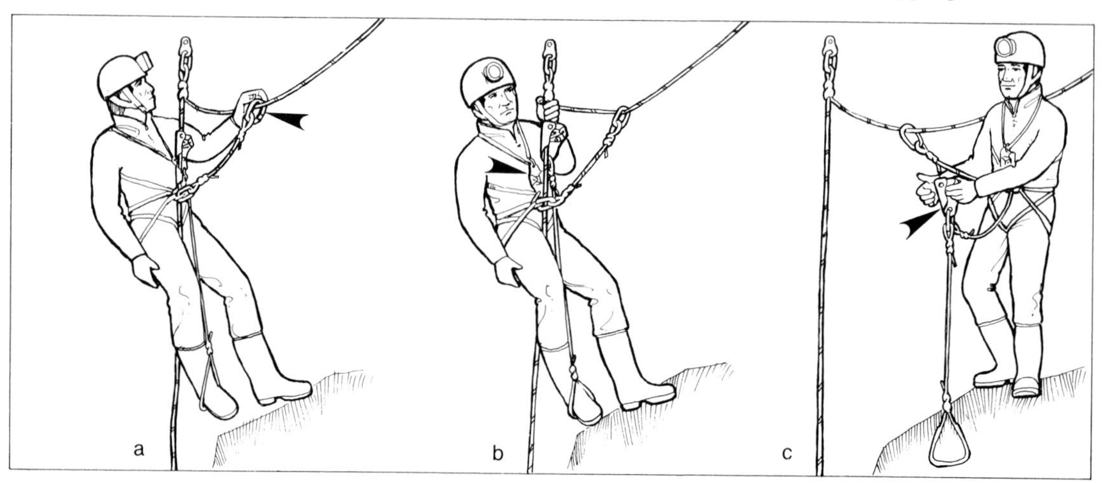

Getting off the rope at the top of an ascent

a. Clip cow's tail into the traverse line

b. Take chest-jammer off the rope

c. Step off the pitch and remove foot-jammer from the rope

Re-belaying

So far, so good. The techniques described will get you safely down, up and off simple pitches without complications. Even in this introductory book, though, it would be wrong to ignore the marvellous invention of the re-belay or *fractionnement*. Cavers in the US, because their technique developed in isolation from rock-climbing, have virtually ignored the re-belay until now, and as a result have had to develop very abrasion-resistant ropes. However, even the toughest rope will wear away when rubbed against limestone, and this should be avoided regardless of the quality of the rope.

Rope re-belaying to avoid abrasion

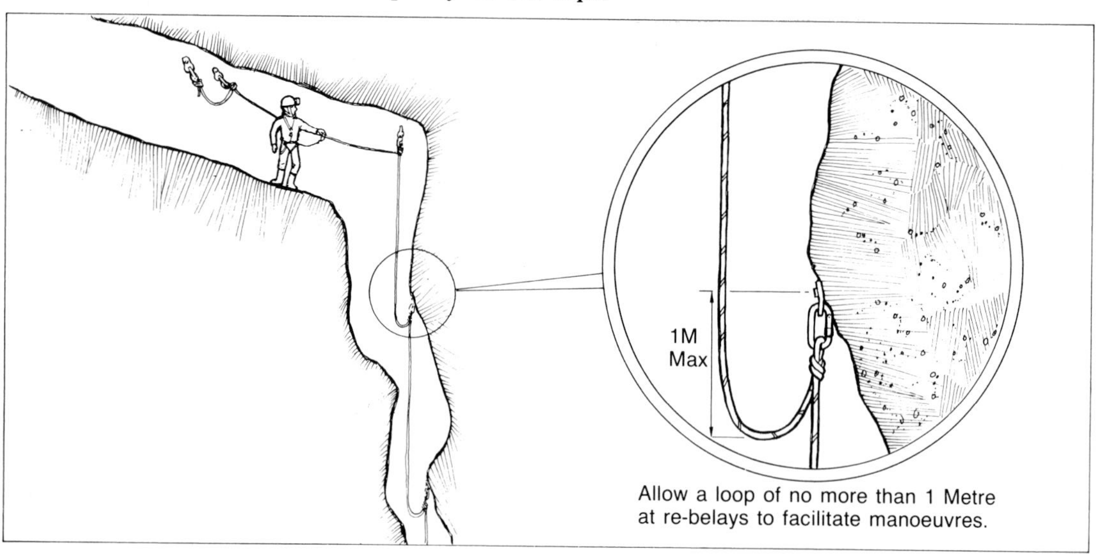

Allow a loop of no more than 1 Metre at re-belays to facilitate manoeuvres.

Re-belaying is the practice of using intermediate rope-anchor points on a pitch to avoid abrasion. On his descent, the first caver looks for possible rubbing points. If there are any, he re-belays the rope at each point, leaving about 0.75m of slack rope each time. Having done this, more manoeuvring is needed to descend and ascend the pitch. The procedure for passing a re-belay on the descent is as shown below:

Passing a re-belay on the descent

a. Clip short cow's tail into re-belay

b. Descend until weight is taken by cow's tail, and, hanging from it, remove descender from rope above re-belay

c. Install descender on lower rope and then remove cow's tail. Unlock descender and continue descent

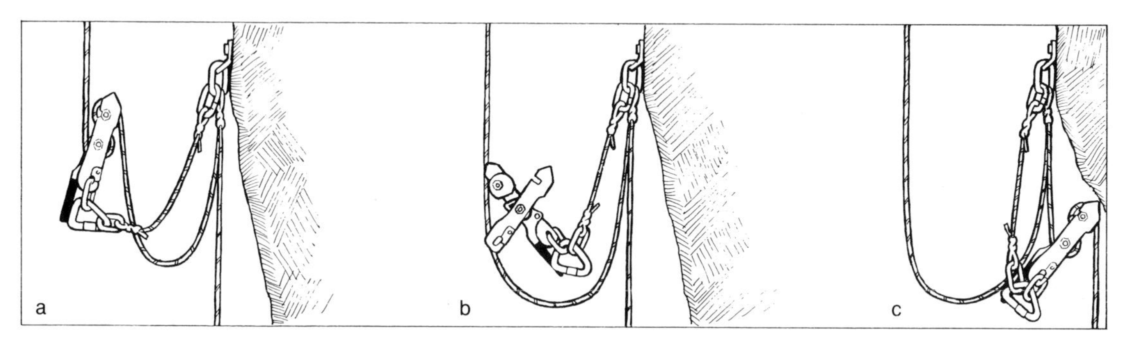

N.B. Only essential gear shown

With practice all this takes only half a minute. The procedure on the ascent is simpler:

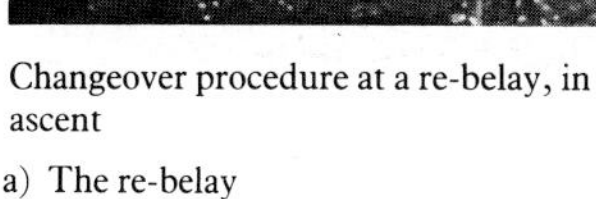

Changeover procedure at a re-belay, in ascent

a) The re-belay

b) Ascending caver clips his long cow's tail into the re-belay

c) Chest jammer is transferred to the rope above the re-belay

d) Foot-loop jammer is transferred to the top rope

e) Unclip cow's tail from re-belay

f) Carry on up the rope

These manoeuvres will suffice for most caving trips. They should be practised a metre or two above ground in a gym or from a tree before doing them 'for real'.

Further techniques

It is useful to be able to deal with the slightly non-standard things that can happen.

On an unknown pitch, it is possible that the rope is too short. (*Always* tie knots in the ends of your ropes before packing them up for underground trips, so that you cannot shoot off the end.) Then you will either have to change from abseil to ascent, or pass the knot and continue downwards on a second rope tied to the first. It is possible to work out ways to do such moves safely, but far better to obtain expert advice, backed up by reading specialist books on vertical caving.

The sit-stand method of climbing ropes has become fairly standardised. Rope-walking, on the other hand, has a number of common variants, but I do not recommend any of them to the beginner. Learn to sit-stand first, and then look at rope-walking only if you intend to tackle a lot of long free-hanging pitches. Although the technique is less tiring on the arms, it is more difficult to get on and off ropes, and considerably more difficult to pass re-belays.

Tie a knot in your rope end.

Ladders

The single-rope techniques we have been considering are taking the place of the older ladder-and-lifeline techniques. In the latter, the caver climbs a flexible ladder, safeguarded by a rope held by a companion. It has to be admitted that this way of doing things is much more sociable! With three or four times the weight of gear to carry, the party has to be larger, and one caver has to help another on pitches. However, in deep potholes the weight becomes prohibitive, and the protection offered by a lifeline is often more imagined than real. The main continuing role for ladders is in the tackling of the occasional short pitch (up to 20m), when the cavers concerned do not want to take, or have not got, an SRT rig. As a beginner you may be introduced to potholing this way.

illustrated p. 108
Ropewalking

Caving ladders have galvanised or stainless wire rope sides, 3–4mm in diameter, with alloy rungs one boot's width wide, spaced at intervals of 250–300mm. Each length of ladder is about 8–10m long, with clip-together C-link fasteners at each end so that any length of ladder can be produced as needed.

Caving ladder, belay wire (tether) and spreader

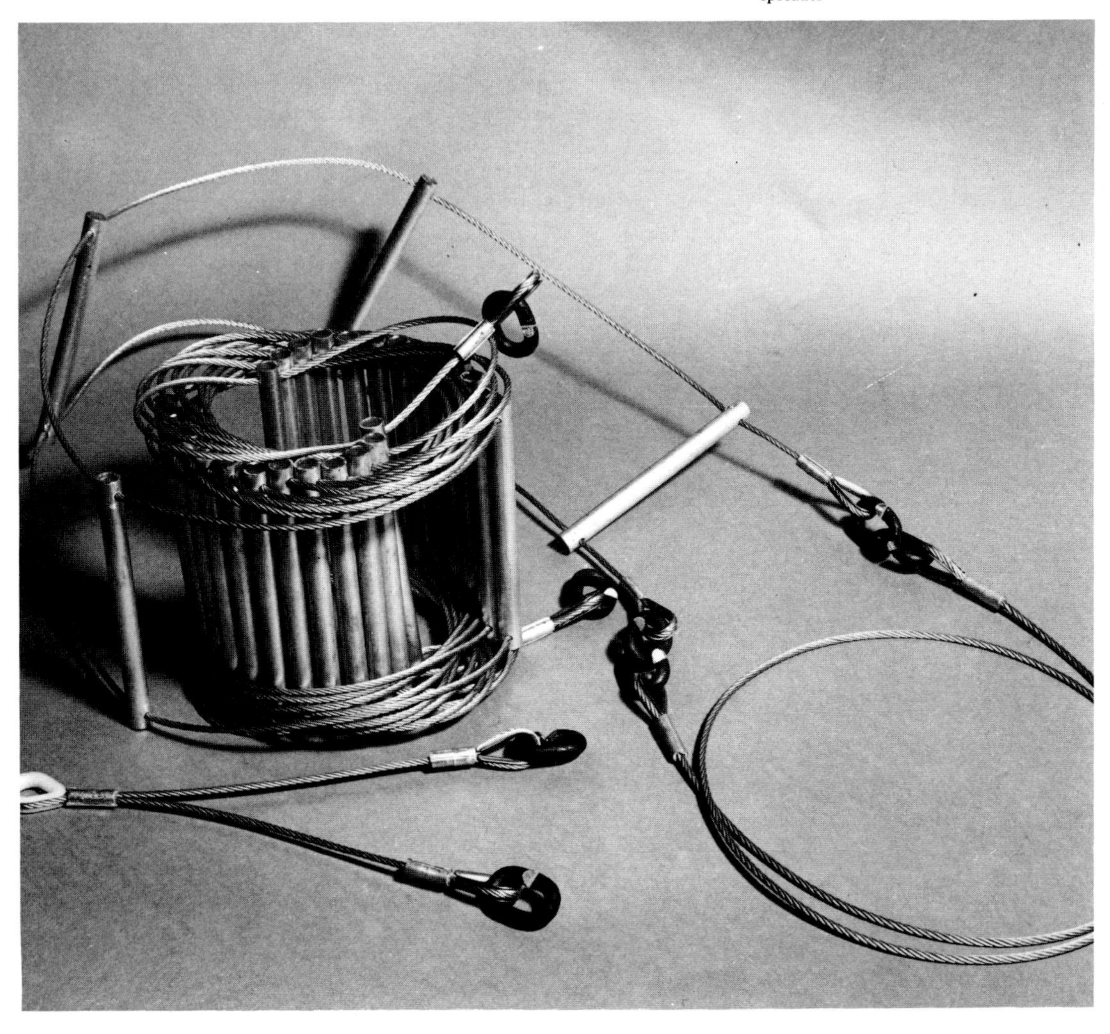

Ladder belayed to a bolt with a spreader

Ladder belayed to a natural 'thread' with a tether

At the top of the pitch the caver who is rigging it secures himself so he cannot slip, and puts a wire-rope 'tether' around a suitable natural belay point or clips a 'spreader' into a bolt belay. As the material of the ladder is harder than the rock, there is no need to avoid abrasion, so the positioning of the ladder is not critical. However, it is useful to hang it so that there are several rungs above the lip of the pitch, to facilitate getting on and off. The required number of ladders are unrolled and clipped together, and the end at the bottom of the pile is connected to the tether. Then the top end is lowered down the pitch.

Skill does help in climbing a flexible ladder. As you hang from it there is a tendency for the ladder to swing away, leaving you dangling like a monkey underneath it. Since you are not a monkey, you knacker yourself in no time at all. Avoid this by wrapping yourself around the ladder as much as possible, with your knees and elbows sticking out, and your hands kept no higher than head level. When climbing directly against a wall it is sometimes useful to swing sideways: this makes it easier to get your feet onto the rungs.

Caver, belayed, lowering the free end of the ladder down the pitch, after securing the top end. Calf Holes, Northern Pennines, UK

Ladders are not as strong as caving ropes. The latter have a breaking load, when new, of 2 tonnes or more, whereas ladders may break at half a tonne or less: any stronger, and they would be too heavy for convenient use. After ageing, there is always a chance that ladders may fail, and in addition even experienced cavers can run out of steam, or simply slip, so ladders should *never* be used without an accompanying safety-rope. Accident statistics bear this out. In the north of England, for instance, there are many short pitches which were usually done on ladders before the 1980s: there is a long list of accidents in the region caused by lack of an effective — or any — safety-rope, even on pitches of less than 10m.

The safety-rope can be of two kinds. First is a fixed rope used for self-lining. This is rigged as before, and used for abseiling down, the ladder being used only on the return. On the ascent the caver, wearing a sit-harness or a complete sit-and-chest harness, tows an ascender up the rope while climbing the ladder, using the rope for resting, or for safety in the event of a fall. To be really safe, it is best to have a chest-ascender in place, so that should a fall occur the climber is left suspended in a sitting position on the rope. He should then have enough extra equipment to carry on up the rope alone.

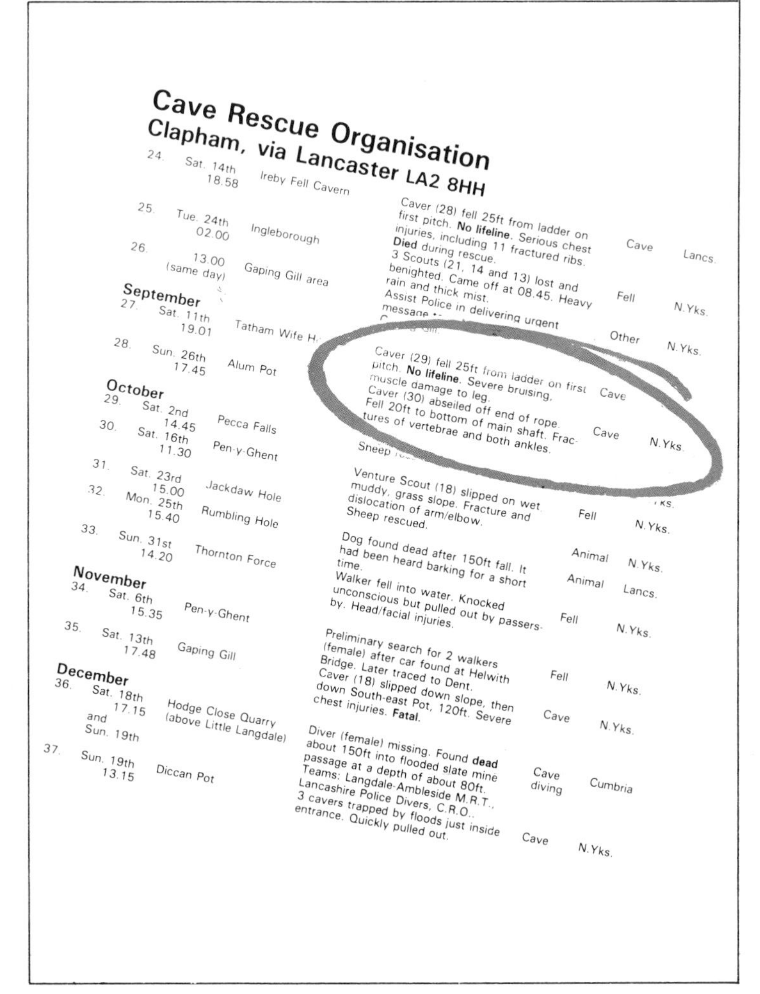

Cave Rescue Organisation
Clapham, via Lancaster LA2 8HH

No.	Date / time	Place	Incident	Type	Area
24.	Sat. 14th 18.58	Ireby Fell Cavern	Caver (28) fell 25ft from ladder on first pitch. **No lifeline.** Serious chest injuries, including 11 fractured ribs. **Died** during rescue.	Cave	Lancs.
25.	Tue. 24th 02.00	Ingleborough	3 Scouts (21, 14 and 13) lost and benighted. Came off at 08.45. Heavy rain and thick mist.	Fell	N.Yks.
26.	13.00 (same day)	Gaping Gill area	Assist Police in delivering urgent message [illegible]	Other	N.Yks.
September					
27.	Sat. 11th 19.01	Tatham Wife H[illegible]	Caver (29) fell 25ft from ladder on first pitch. **No lifeline.** Severe bruising, muscle damage to leg.	Cave	
28.	Sun. 26th 17.45	Alum Pot	Caver (30) abseiled off end of rope. Fell 20ft to bottom of main shaft. Fractures of vertebrae and both ankles.	Cave	N.Yks.
October					
29.	Sat. 2nd 14.45	Pecca Falls	Sheep [illegible]		[illegible]ks.
30.	Sat. 16th 11.30	Pen-y-Ghent	Venture Scout (18) slipped on wet muddy, grass slope. Fracture and dislocation of arm/elbow.	Fell	N.Yks.
31.	Sat. 23rd 15.00	Jackdaw Hole	Sheep rescued.	Animal	N.Yks.
32.	Mon. 25th 15.40	Rumbling Hole	Dog found dead after 150ft fall. It had been heard barking for a short time.	Animal	Lancs.
33.	Sun. 31st 14.20	Thornton Force	Walker fell into water. Knocked unconscious but pulled out by passers-by. Head/facial injuries.	Fell	N.Yks.
November					
34.	Sat. 6th 15.35	Pen-y-Ghent	Preliminary search for 2 walkers (female) after car found at Helwith Bridge. Later traced to Dent.	Fell	N.Yks.
35.	Sat. 13th 17.48	Gaping Gill	Caver (18) slipped down slope, then down South-east Pot, 120ft. Severe chest injuries. **Fatal.**	Cave	N.Yks.
December					
36.	Sat. 18th 17.15 and Sun. 19th	Hodge Close Quarry (above Little Langdale)	Diver (female) missing. Found **dead** about 150ft into flooded slate mine passage at a depth of about 80ft. Teams: Langdale-Ambleside M.R.T., Lancashire Police Divers, C.R.O..	Cave diving	Cumbria
37.	Sun. 19th 13.15	Diccan Pot	3 cavers trapped by floods just inside entrance. Quickly pulled out.	Cave	N.Yks.

Self-lining is not often needed, but it can be very useful. Two instances are (a) when a rope cannot be rigged to avoid all abrasion, but it is reasonable to abseil carefully (but not to make a jerky ascent of the rope); and (b) when the lifelining technique described below is to be used by all except the party leader. A similar technique is used if a rope ascent is safeguarded by a second rope, which has a spare jammer towed up it in case the main rope fails. Twin-rope technique is widely used in the Eastern bloc.

Climbing a ladder

a) Keeping one's centre of gravity as close as possible to the ladder reduces the load on the arms — good!

b) A very strenuous way to climb a ladder — bad!

Self-lining. If the climber falls, he is held on the fixed rope by his jammer

Lifelining

If the ladder is to be descended as well as ascended, 'traditional' lifelining is called for. The lifeliner belays himself at the head of the pitch. Ideally he should be able to see down the pitch, but with the rope tight between him and his belay he must be sure not to fall over the edge. He fixes a separate belay next to himself into which he can clip a karabiner, preferably at about shoulder height. The caver who will climb down the ladder ties a figure of eight loop in the free end of the lifeline rope and clips this to his load-bearing belt or harness with a screw-gate karabiner. The lifeliner then makes a sliding friction knot which runs through the separate belay just mentioned. There are two such arrangements commonly used — the Italian hitch, and the sticht plate. The former knot is used in conjunction with a large, pear-shaped karabiner; the latter is a metal plate with a hole in it through which a loop of rope is pushed, and the loop clipped into a karabiner. (This indirect form of belaying has superseded body-belaying, where the lifeliner held the rope around his body. This may still be seen in action, but should not be copied.)

The Italian hitch works thus: a double loop is made as illustrated, and clipped into the special karabiner. The rope can now be pulled through the knot in either direction, reversing itself through the karabiner as the direction of pull is reversed. The rope is fed through the Italian hitch as the caver descends. Should he need support, or even fall off, the lifeliner can hold the climber without difficulty by gripping the rope as shown. On the ascent, with the knot reversed, it is easy to feed the lifeline back through it, and to stop a fall if need be. The lifeliner must ensure that he keeps up with the movement of the ladder climber, and that he keeps permanent hold on the rope on the other side of the knot from the climber. The friction introduced into the system by the hitch allows a competent small person to hold a larger person without strain, and to lower them down if necessary.

The sticht plate works in a similar fashion, and is commonly used by rock climbers.

Lifelining
a) Caver and Italian hitch attached to belay point

b) Caver belayed and Italian hitch clipped to himself. NB The belay point should be 2 metres higher, but this photo illustrates the sort of practical difficulty that arises in real life

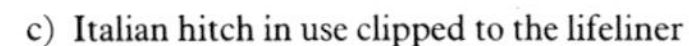

c) Italian hitch in use clipped to the lifeliner

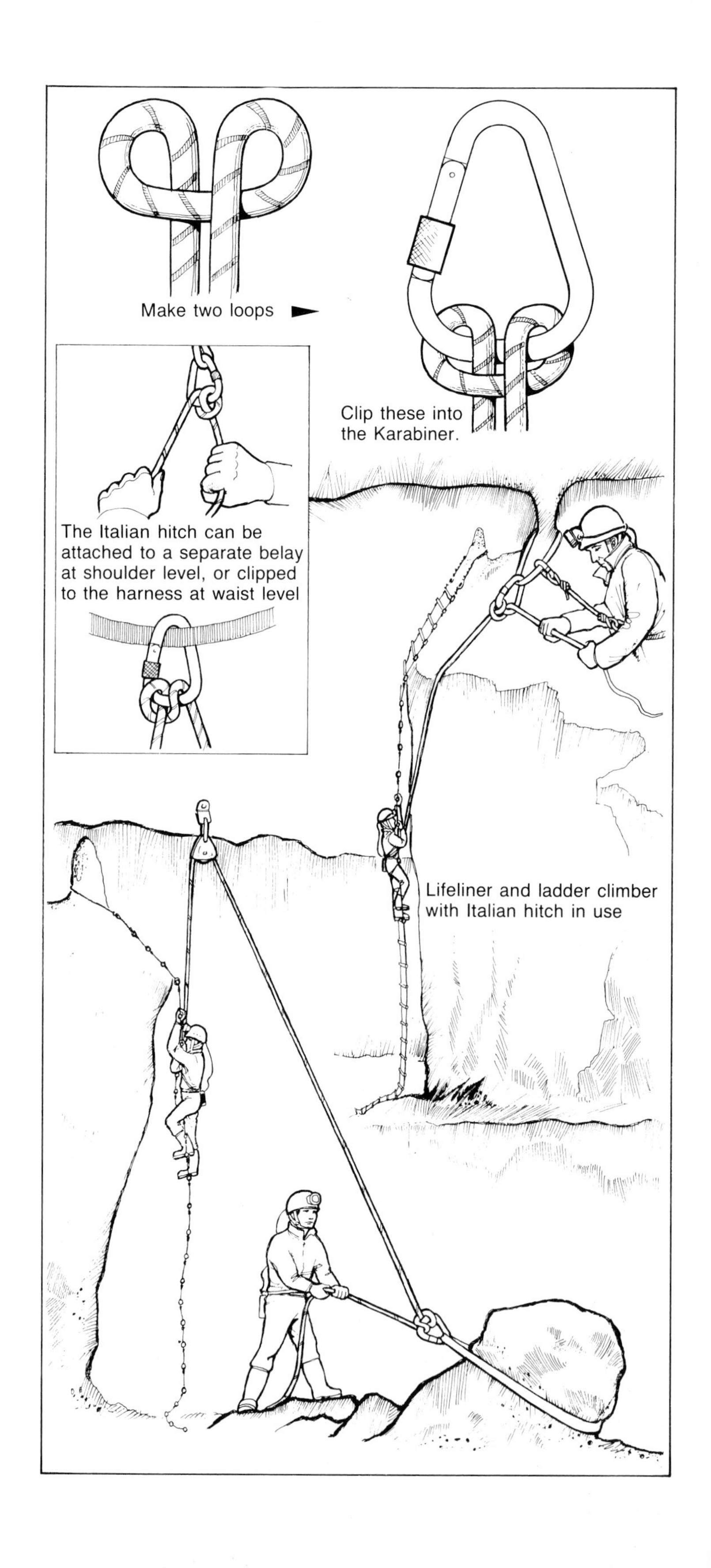
Make two loops
Clip these into the Karabiner.
The Italian hitch can be attached to a separate belay at shoulder level, or clipped to the harness at waist level
Lifeliner and ladder climber with Italian hitch in use

The Italian hitch and lifelining

These devices require exactly the same sort of rope as for abseiling and prusiking. However, the constant friction from the rubbing they receive will gradually wear them out fairly evenly along their length, so it is best to allocate separate ropes to lifelining, and replace them when they become 'furred up'.

Signalling between lifeliner and climber can be difficult — this is one very good reason for confining use of the technique to moderate pitches. Misunderstood calls have caused deaths on pitches as short as 16m. The code to use, preferably with a whistle, is:

- One blast: stop.
- Two blasts: coming up (take in).
- Three blasts: going down (let out).
- Four blasts: all OK.

Verbal signals are like those used in climbing. With the lifeliner at the top of the pitch, and the ladder climber at the bottom wanting to come up, the sequence is as follows:

- ***Climber:*** 'Take in.'
- ***Lifeliner:*** 'Taking in.'
- ***Climber*** (when all the slack rope has been taken in): 'That's me.'
- ***Lifeliner*** (when he is sorted out): 'Climb when you're ready.'
- ***Climber:*** 'Climbing.'
- ***Lifeliner:*** 'OK.'

Use of a rope whose length is the same as that of the pitch means that the lifeliner must be left at the top, unless he uses abseil and self-line techniques. However, if the rope is double the length of the pitch, everyone can be lifelined. The last man, having lined the rest of the party down the pitch, clips the rope running from himself through a pulley which is attached to a sound belay positioned so that the rope will not snag. One of the party at the bottom then fixes a belay to take an *upward* pull, and lines the last man down the ladder. Before the party moves off, the two rope-ends are tied together, to make sure that the rope cannot run out of the pulley. On the ascent the process is reversed.

Conclusion

Vertical caving is becoming something of a sport in its own right. Ascending races, long a feature of cavers' gatherings in the USA, are starting to appear in Europe. Fair enough, so long as the caves do not suffer.

The real justification for all the gear and techniques we have talked about is that they make the exploration of the underworld possible, and reasonably safe. Fifty years ago, in the era of wooden-runged ladders, roughly 8kg of gear was needed for every 10m of pothole descent. Nylon ropes and metal ladders reduced this to 4kg. Now, with single ropes, the figure can be reduced to less than one kilogram. Descents that required expedition-type assaults can now be done by two sorted-out cavers in a weekend!

Check-list of personal gear required for vertical caving

1	Foot-loop jammer	Standard: Alternatives:	Petzl jammer Petzl, CMI or Jumar handled jammers
2	Linking Maillon/ karabiner	Standard: Alternatives:	7mm steel Maillon Rapide Small oval steel or alloy screw-gate karabiner
3	Foot loop	Standard:	Low-stretch foot loop
4	Security-link	Standard:	Dynamic climbing-rope strop
5	Chest-harness	Standard: Alternative:	Figure-of-eight chest-harness strop 'Classic' chest harness
6	Chest jammer	Standard:	Petzl Croll
7	Main attachment	Standard: Alternative:	Steel 10mm delta Maillon Rapide Alloy 10mm delta Maillon Rapide
8	Sit-harness	Standard: Alternatives:	Standard sit-harness 'Rapide' sit-harness, 'Avanti' sit-harness, belt and leg loops
9	Descender	Standard: Alternatives:	Petzl 'Stop' descender Petzl single descender, Rack, Lewis self-locking descender
10	Cow's tails	Standard:	Dynamic climbing rope twin strop
11	Karabiners	Standard:	Small oval alloy or steel karabiners
12	Personal equipment bag		

A lightweight and extremely versatile jammer that can be used also for tackle-hauling and self-rescue
Some prefer to hold a handle; however, this can lead to poor technique

A cheap, lightweight and very strong attachment link
A karabiner can be detached easily, and if necessary be put to other uses

A single foot loop made of low-stretch rope – 8mm terylene

A climbing-type rope of 9mm diameter which can sustain a drop with a high fall factor

A quick-adjust chest strap that holds the chest jammer efficiently, without constriction
'Belt and braces' type harness, favoured by some

The Petzl Croll is specifically designed for chest mounting

An extremely strong and inexpensive link that connects chest jammer and sit-harness
A lighter-weight version of the steel Maillon (NB: *Never* use a Krab in this position)

A supremely comfortable harness made from 50mm webbing, with waist bands and thigh loops which are fully and independently adjustable
The 'Rapide' is easy to put on and adjust; undoing the sit-strap allows unrestricted movement along horizontal passages.
The 'Avanti' is a lightweight but effective sit-harness, and is fully adjustable

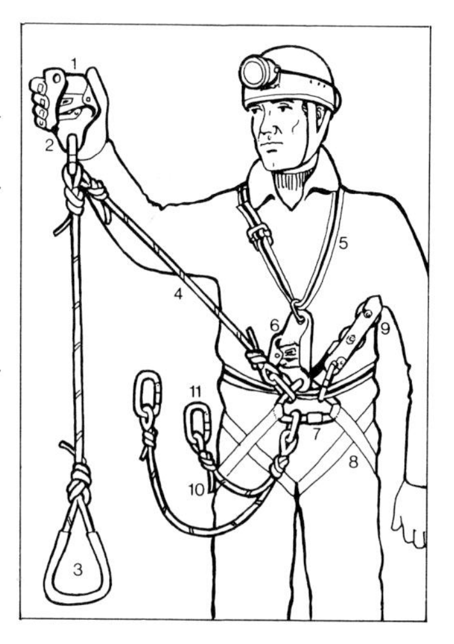

A reliable and compact descender that instantly locks onto the rope once the stopping handle is released
The Petzl is a compact single descender that does not twist the rope during descent; *not self-lock*. The Rack can be used with ropes of varying widths and double ropes; *not self-lock*.
The Lewis can be used with double ropes

2.5m of 11mm climbing rope that can sustain a drop having a high fall factor, and minimise the shock-load on both belay and caver

Approximately six karabiners are required in a full SRT kit. Light alloy karabiners are OK for all uses except where they may be subject to wear, such as the friction karabiner while descending

A simple bag clipped to the waist to hold all your gear

A cautionary note

Caving is not as dangerous as popular opinion has it. Perhaps unfortunately, every underground incident is news, yet activities with a far worse accident record, such as climbing, motorcycling and skiing, do not attract such intense and morbid press interest. In fact, if due care is taken, caving is one of the safest of the so-called 'risk sports'. The good caver sets out not to conquer the cave but to explore it safely, using whatever equipment is necessary to do so, and avoiding hazard to himself and damage to the cave.

As a caver, it is essential that you understand the potential hazards of caves: it is this understanding which prevents a possible danger becoming a real one. Develop the attitude of mind that weighs up the risks of a cave both beforehand and as you go through it. Be fearful if necessary. Fearlessness can lead to foolhardiness, and a strong yellow streak is very good for survival!

In this chapter we shall look at caving's potential hazards.

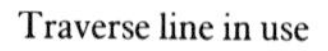
Traverse line in use

Falls

Beware of those minor slips which can cause a twisted ankle or a broken wrist. Climb down even small drops — *never* jump — and tread carefully at all times. I have seen a broken knee-cap resulting from a one-metre fall! Getting a casualty out of a cave can be very difficult, even with a minor injury. Beware of black holes in the floor. 'Out of sight, out of mind' is an easy but dangerous attitude to develop. In the darkness great drops are less frightening, and are often treated too lightly. Always clip or tie into a rope or belay at the head of a pitch. Be especially careful if the rock is loose or muddy, and remember that cave floors are usually very hard. A fall of 10 m or even less will often be fatal underground.

Beware of poor equipment, and poor use of equipment. There is no excuse for accidents caused by bad gear or bad techniques; they are entirely preventable. If you are not satisfied that your pitch-rigging is safe, do not go down. Remember that two doubtful belays do not equal one safe one, and that it is too late to protect your rope after it has frayed through. If you are using ladders, remember that sloppy lifelining may be even more dangerous than no lifeline (it may endanger the lifeliner also), and no lifeline is an underground form of Russian roulette.

Beware, too, of things falling onto you. Dodging flying missiles at the foot of a pitch is a silly game. If pitch-tops and ledges are cleared as the pitch is rigged, this danger is minimised, but it is still wise, if practicable, to keep well clear of anyone above. Never throw stones down a shaft unless you are certain that there are no cavers and no tackle down it (even then you should only do this for the purpose of gauging the depth of an unknown shaft). You

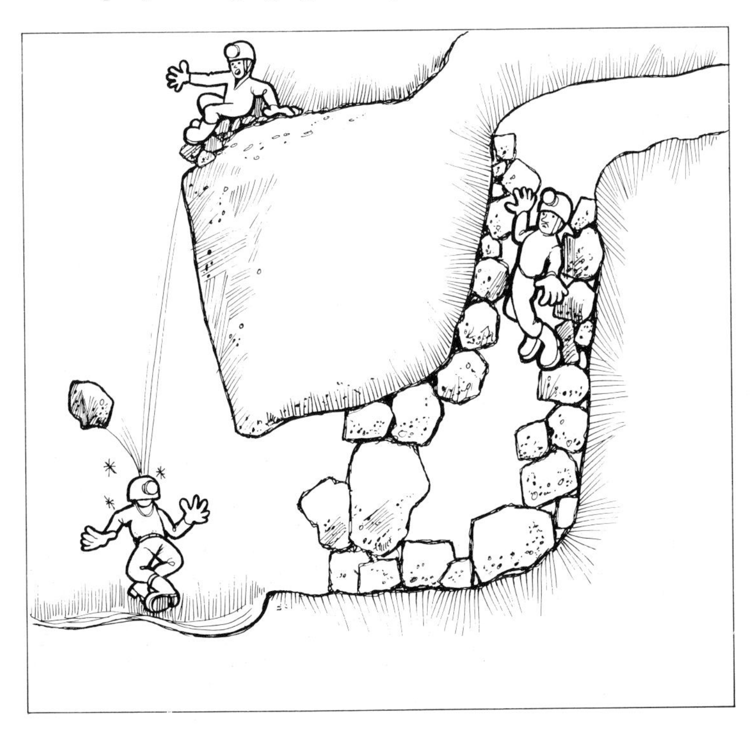

can avoid bringing loose passages down around you by moving more slowly and carefully than the caver in front: then if he does not cop it, you are unlikely to! Loose material is less likely in well-trodden caves, but it can be an ever-present hazard in 'new' caves, which have more 'hanging death' waiting to be disturbed.

Current thinking is moving towards the idea that caving helmets should all be of climbing-helmet standard, in order to give some protection against falls. Miner's and construction-type helmets really protect you only against bumping the ceiling with your head.

Water

Wet-cold conditions and inadequate clothes are not good

The hazards presented by water are more subtle than those offered by falls, which are fairly obvious if you think about them a bit. Apart from earthquakes, and the odd unstable boulder-choke, the fall hazards of a cave remain fairly constant. Water hazards are more fickle, varying with the amount of water in the cave, and its temperature. We must examine the way in which they do vary.

Most caves occur in hilly or mountainous country. In such terrain there is only thin soil cover, steep slopes and relatively little vegetation to intercept rainfall or transpire it. There is little to impede the passage of the rain. Upland streams formed on catchments of impervious rock, boulder clay or peat are subject to rapid rise in level after the onset of rainfall or snow-thaw, and this soon reaches any cave in the stream's path. If the surface above the cave is limestone it will often have no vegetation or soil at all, so the rain runs freely down every crack, and percolates directly through to underground waterways.

Precisely because of this rapid run-off, cave streamways, like the hillside streams above ground, dry to mere trickles in fine weather, filtering slowly along, or even dry up altogether. In these conditions the only significant water underground is in pools, canals and sumps, with no appreciable flow. Progress is easy, and you may get into parts of a cave which in bad weather are flooded.

If it rains or thaws while you are underground, the build-up of water is liable to be fast and sudden. A stream which has a gradually rising input of water at the top end does *not* rise gradually further down. Instead, the rising flow overtakes the existing trickle, is slowed down by the friction of the stream-bed, and is itself overtaken by even more water from behind. The result is a 'flood pulse'. At its most dramatic, this can mean that one minute there is no water in a passage, but the next it is a raging torrent. Unlike surface streams, underground streams have no banks to retire to if this happens!

A feature of caves in flood is that normally dry passages, above the level of permanent streamways, can become flood overflow channels with no warning — other than a possible change in wind-draught — at all. If the flood meets a constriction — a boulder-choke, for instance — it will 'back up', and can sometimes fill the cave behind to amazing depths. The most astounding one is the Grotte de la Luire, in France. In dry

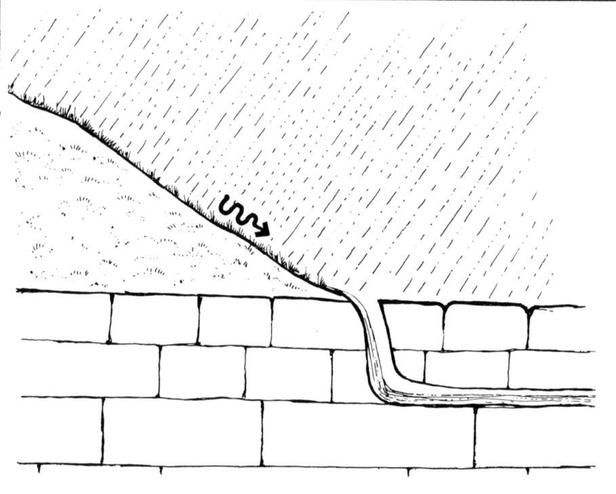

Rapid runoff occurs on steep, sparsely vegetated slopes soon after the onset of rainfall:

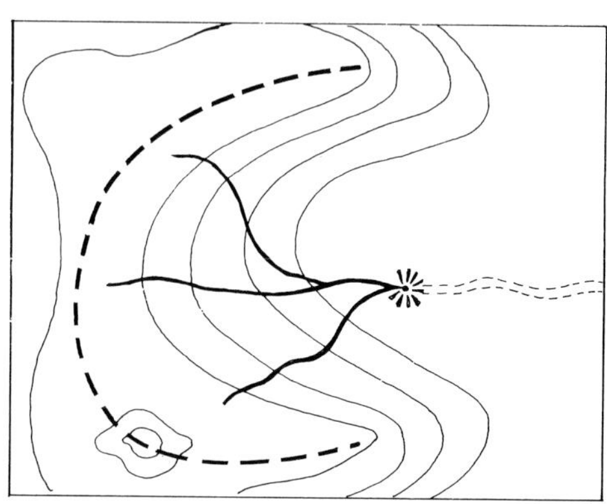

Upland catchments draining into cave systems are often approximately semi-circular. Small streamlets are focused towards the cave entrance and will all rise at the same time.

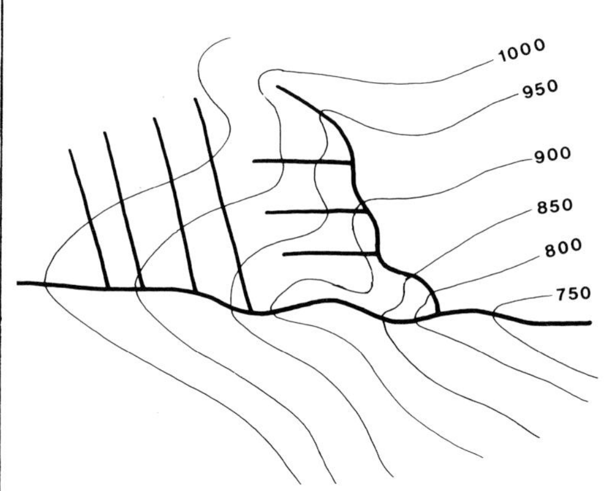

Gripping — the draining of upland by frequent trenches speeds the reaction to rainfall ....

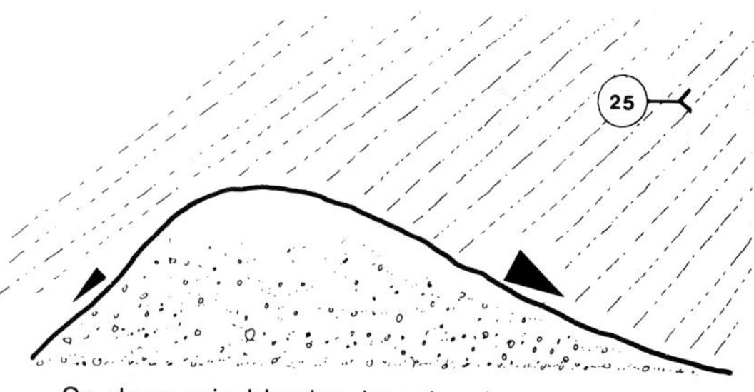

... So does rain blowing into the slope

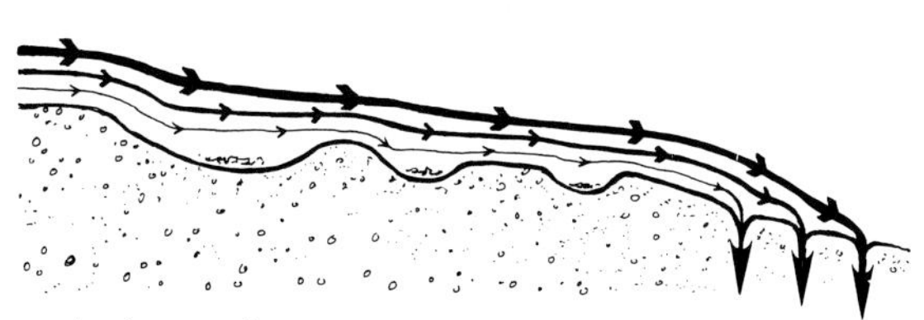

In dry weather water trickles along stream beds, or disappears entirely through cracks in the limestone. After the onset of rain, water starts to feed into the upper reaches of the bed. This flows faster, overtaking any residual flow, but is then slowed as it hits 'dry' stream bed. Continued flow at the 'top end' flows over the initial flow, building up a flood-pulse.

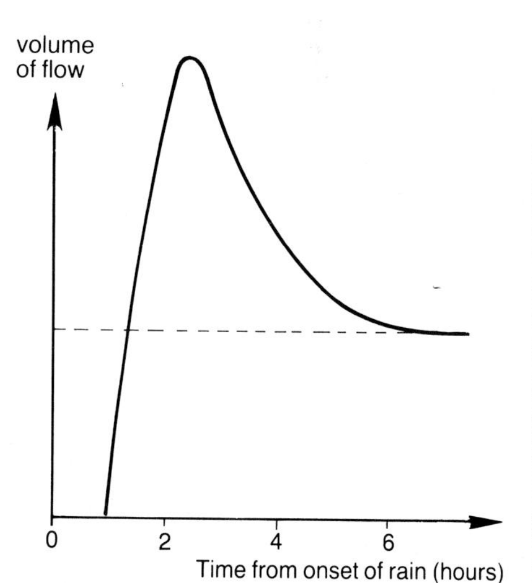

Typical flood hydrograph for a cave-stream after the onset of steady rain.
No flow is transformed very quickly, and with no warning, into a peak, which falls more slowly.

weather it can be descended for more than 500m vertically, but in exceptionally wet weather flood-water pours from its entrance! There are a number of caves which flood completely like this, and which should never be attempted in unsettled weather conditions.

Even with great experience it is not easy to tell how long it will take a cave to flood after the onset of rain. It depends on the size and type of catchment draining into the cave, the direction and intensity of rainfall, and the soil-moisture or snow conditions at the time. Generally speaking, the larger the catchment (drainage) area, the slower the flood; the harder the rain and the soggier the ground to start with, the faster the flood. If rain is being blown onto a slope, it will catch more rain than the lee slope.

Nearly all cave systems which respond to rainfall will do so a few hours or *less* after the onset of heavy rain (or sudden snow-thaw). Persistent heavy rain falls at a rate of around 4mm per hour (this is a typical rate of fall for the warm fronts brought to Britain from the Atlantic by the familiar depressions): it is this rate of rainfall which is likely to result in caves flooding in one to four hours. Fortunately, depressions are usually fairly accurately forecasted, so trouble can be avoided.

More dangerous are summer thunderstorms and spring thaws. The former tend to occur towards the end of a period of high pressure. Because the weather has been fine, water levels are low, and conditions underground are ideal for exploration. The day dawns clear and fine, and the caving trip begins. Thermals develop, and the first rumblings of thunder are heard around the start of the afternoon. Any time after this the heavens open. Typical thunderstorm-rain intensities are around 100mm per hour, or 25 times as heavy as 'normal' heavy rain. The heavier the rain, the quicker the flood, so that after the onset of a heavy thunderstorm it will be a matter of minutes, not hours, before caves flood. The flood levels may be catastrophic.

The lower Long Churn streamway in normal, and flood, conditions. Northern Pennines, UK

Check the flood risks of your chosen cave before visiting it by reading guidebooks and, if possible, obtaining local knowledge. If there is any flood risk, obtain a weather forecast: do this as near to the time of your trip as possible — it is no use checking on Monday to see if conditions will be all right for the following weekend. Find out if surface-stream levels are acceptable for you to do the exploration, and check the weather forecast on the *same day* as your trip.

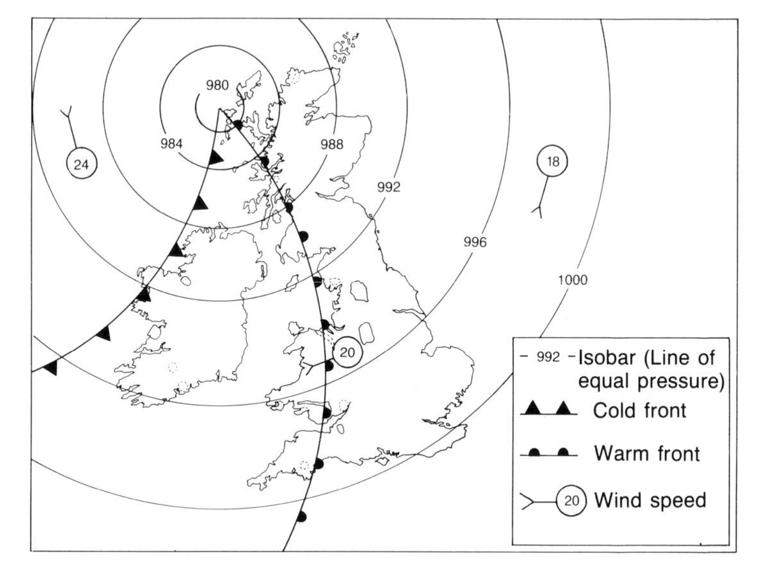

Typical westerly depression over the British Isles. The effects of the warm front have reached the main caving areas — with several hours of continuous rain. This will be followed by a cold front and showery weather. With this forecast it would be foolhardy to enter a flood-prone cave

The experienced caver should be able to tell whether or not a cave floods by the presence or absence of delicate formations, mud and tide-marks on walls, the presence of washed-in vegetation and other signs. Evaluating such evidence is not always straightforward. If a passage floods violently, it is likely to be washed clean; if it 'backs up' behind a constriction, it will probably be mud-covered. Some caves flood completely, some have permanently dry high-level routes, and others never flood at all. If in doubt, don't risk it. Flood hazards have been consistently underestimated by cavers down the years.

Evidence of flooding: in the passage roof, Kingsdale Master Cave, Northern Pennines, UK

Evidence of flooding: in Calf Holes, Northern Pennines, UK

SLEETS GILL CAVE NGR SD. 959693 Grade IV
Alt. 950 ft. (290 m.) Length 1½ miles (2.4 km.)
Depth 90 ft. (27 m.) Vertical range 250 ft. (76 m.)
Explored 1906; extended 1968, U.L.S.A. and B.S.A.; 1971, N.C.C.; 1972, U.L.S.A.
WARNING—Entrance slope can sump rapidly in wet weather, often after long delay. The whole known cave except The Ramp floods to the roof.
One of the most spectacular caves in the district. Entrance is at the head of Sleets Gill near Hawkswick, an arch 5 ft. (1.5 m.) high and 12 ft. (3.7 m.)

A short, wet pitch in Tatham Wife Hole, Yorkshire, UK. If the stream rises while the caver is below he will be in trouble on his return!

If you are faced with a sudden rush of water through the passage you are in, you will have to make the best decision you can in the circumstances. In general (though not always), the best advice is to climb to the highest place you can find and sit it out until the flood subsides. Use your emergency bag, huddle together, conserve your food, keep well away from draughty places such as the bottoms of pitches, and use as little energy as possible. You may need it later! Wet pitches can soon become unclimbable, even though the passages below and above are negotiable.

Prevention is better than cure in this case, so whenever possible pitches should be rigged away from potential flood water. This is usually done by starting the pitch with a traverse which will take the 'hang' beyond, or round a corner from, the falling water. Flooded pitches can cause more than a temporary setback: floods often carry stones and even boulders with them, and can thus damage or break gear hung down waterfalls.

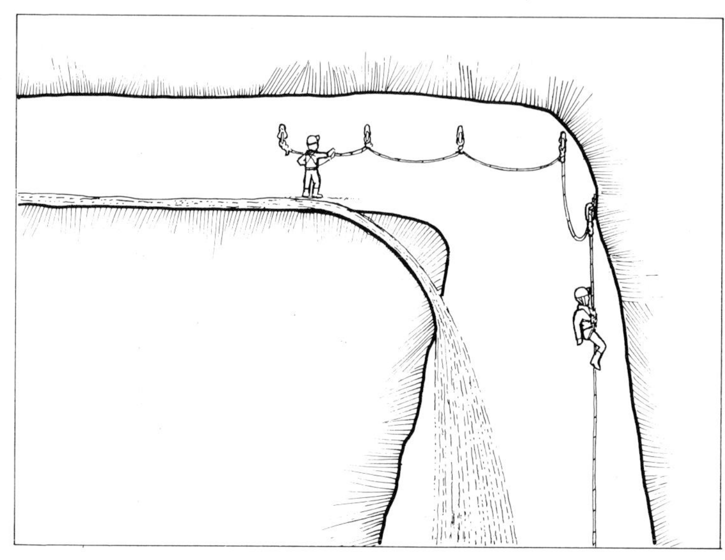

Pitch rigging to avoid water. Traverse over and beyond the problem!

The hazards of deep water and sumps have already been covered in Chapter 5 (see pages 78-79), but perhaps it is worth pausing here to explain just why sump-diving is potentially so hazardous. Unlike open water, the way out of a sump is not straight up, but along — usually involving swimming back through mud stirred up on the way in. With zero visibility and bulk-adding air bottles, the cave-diver has a difficult task, even with the guideline laid through the sump. If he loses the line, or gets jammed, or his air-valve malfunctions, he is in big trouble. The only way to cave-dive in reasonable safety is to join a Cave Diving Group, and to learn from its experienced members.

Swimming into a sump is relatively easy. Visibility on the return may be nil because of stirred-up mud

Abseil on
this rope
Then pull
rope down
by hauling
on this one
How to abseil and pull the rope down after you

Water in caves is often cold, sometimes very cold. An unclothed person would die in a matter of minutes in melt-water, and it is this sort of temperature you may experience in some underground stream passages. Adequate clothing to protect against the effects of such cold water is essential. Without it, the onset of exposure — when the body loses heat more rapidly than the metabolism can generate it — can be very rapid. A useful tip in cold caves is to remember that a lot of body heat-loss is from your hands, feet and head. Adding a hood and gloves to good body-protecting clothes makes a big difference. Adequate protection against wet-cold is particularly important if you are injured or stuck underground. Avoiding exposure is much easier than dealing with its effects.

Vladimir Ilyukhin, leading Soviet speleologist, killed in a cave accident in 1982

Mines

Exploration of disused mines is a leisure activity in its own right, but one which frequently overlaps with 'natural' caving — as for instance in Derbyshire, where numerous caves have been found through the work of miners, and most underground trips are a mixture of cave and mine exploration. Mines can have all the hazards of caves, and they possess additional ones of their own. Apparently solid floors may simply be rubble supported by rotting timbers. Shoring considered essential by the miners decades ago may now have disintegrated to leave various forms of hanging death. Air circulation in mines tends to be poorer than in caves, so that 'bad air' is a greater problem (added to by gases being given off by rotting timbers and the rock itself). All the advice about caving with experienced guidance applies even more strongly to old mines.

Getting stuck

This is not easily done, because common sense stops most people pressing on until return is impossible. Nevertheless it can happen. Cavers digging through choked passages have to be particularly careful not to be trapped by roof-fall. In tight horizontal crawls, equipment sometimes snags up under your

Old mine passages can be less than safe!

body. Where this could happen, it is best to remove such equipment from your body and push it ahead of you (helmets and electric-battery cases can fall into this category). Avoid round pebbles rolling under your chest by brushing them out of the way first.

Horizontal squeezes tend not to be man-traps, but vertical ones can be more serious. It is more difficult to climb out of a tight vertical slot than to slide into it; so, if there is no alternative way back, be careful. At least one caver who slid down into a cul-de-sac is still there. A sobering thought; but getting physically stuck is a minimal hazard for the fit caver.

It is, however, a major problem on cave rescues. Passages which are easy to negotiate with an active, flexible body can become impossible for someone strapped to a rigid stretcher — a very good reason for avoiding injury.

There is another way of getting stuck. 'Through trips' — going in at the top entrance of a cave system and coming out at the bottom, descending the pitches using a doubled rope, and pulling it after you — are tremendous fun. Not as strenuous as having to climb back up pitches, with different surroundings all the way, they have great attraction. However, if a rope should get stuck, or if you should take the wrong route so that the whole party lands up at the bottom of a cul-de-sac, you are in trouble!

Prevention is better than cure. Check all ropes *before* the last man abseils down to see that they will run, and do not attempt complex through trips without a guide who knows the route. I must add that, since the rope has to be pulled down the pitch, belaying tends to be done from only one point. This is bad practice: at least two belays should be linked by a short length of rope, which you subsequently abandon. Do not trust rope loops left behind by others; always replace them with your own.

Getting lost

Experienced cavers rarely get lost. The novice may well credit them with supernatural route-finding ability, but it is actually just a matter of applied common sense. Do not try to remember the entire cave, just the junctions. Look back after each junction to note what it looks like when going the other way.

If passage marking for the return is unavoidable, leave small tape or cloth arrows, or put one stone on another. (Remove on your way out)

Following streams is easy when you are going downstream but, going upstream, every junction presents a choice of route. Hence, if you find yourself exploring upstream there is no need to take note of any splits in the stream. On your return all routes lead downstream.

If the dip of the limestone beds is obvious, this can be used as a guide. In maze-caves a compass is useful. Strong draughts can be a useful pointer, especially in deep systems. If you have to leave markers, remember that they must not damage the cave and should be removed on return. One pebble balanced on a rock will indicate the right turning, or you can take small strips of fluorescent material.

If you do get lost, never wander on regardless, but try to retrace your steps until you recognise the passage, and do your best to find the right way out again. If you have to wait for rescue, do so in the biggest and most obvious place you can find.

Small parties are particularly vulnerable. If they take longer than they intend — perhaps by getting temporarily lost — they tend to have less light in reserve than would a large party. Most cases involving parties 'lost' down a cave turn out to be a case of 'no light left'!

Problems of timing and numbers

Having lots of people in a party may make a trip last much longer than anticipated, due to waits at tops and bottoms of pitches, and hold-ups at bottlenecks. Big parties have occasionally lost one of their number simply because nobody knew how many were supposed to come out! It is thus a good idea, when possible, to stick to fairly small groups — four is probably ideal. This gives enough carrying power for most trips, but cuts down the waiting about. If a mishap should occur, there will be someone to stay with the casualty, and two to go out for help. (Always send two people if possible: one person rushing out of a cave alone is liable to come to grief.)

Experienced cavers often cave in pairs only, or even solo. This is not recommended until you are very experienced and capable of weighing up the extra risks involved.

Before going underground, the leader of a party should leave details of what is to be attempted, and the estimated time of

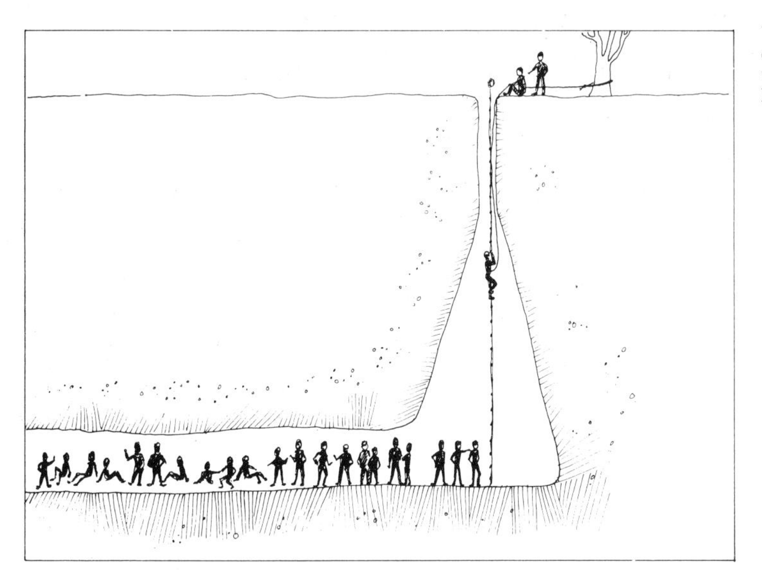

The bottleneck effect. If it takes the first caver ten minutes to climb out of the cave, the Nth caver will not reach the surface for N × 10 minutes!

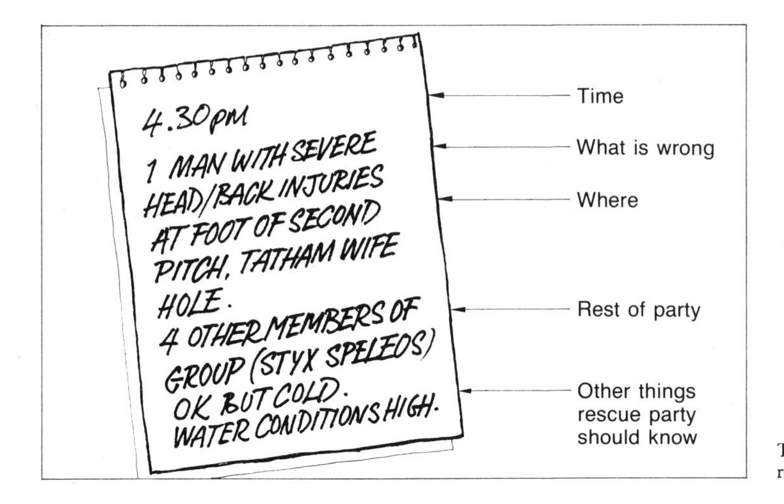

4.30pm
1 MAN WITH SEVERE HEAD/BACK INJURIES AT FOOT OF SECOND PITCH, TATHAM WIFE HOLE.
4 OTHER MEMBERS OF GROUP (STYX SPELEOS) OK BUT COLD.
WATER CONDITIONS HIGH.
Time
What is wrong
Where
Rest of party
Other things rescue party should know

The information you need to include in a rescue message

return. This can be telephoned through to someone reliable back home, or left with someone in the caving district. The person left with the responsibility for raising the alarm should also be told just what to do if the party does not report back. Leaving a reasonable time-margin for unforeseen hold-ups will save a lot of false alarms. The importance of letting someone know where you are was brought home to me at Easter 1972. Some local lads called on me one afternoon to say that three bikes had been parked near the entrance to some local caves since the morning of the day before. On investigating we found three boys huddled, in the dark, in a flood-prone passage, a long way from daylight. They had been on the first day of a week-long youth-hostelling trip when they entered Ibbeth Peril Cave, with enough carbide in their acetylene lamps to last an hour. Getting lost, their lights quickly expired, and but for the chance finding of their bikes they too would have expired!

Ogof Ffynnon Ddu. Streamway in normal and flood conditions

illustrated p.130
Water resurging from the normally dry entrance to the Grotte Bournillon, France. Note the figure in the distance!

Bad air

Most caves have good air-circulation; indeed, because of their pure air, caves have been used as a therapeutic aid for patients with lung illnesses. There are exceptions, most of them due to the build-up of carbon dioxide. Best known bad-air caves are those of Bungonia in New South Wales, Australia, whose air has a carbon-dioxide content many times greater than normal. In the UK, there have been some problems of carbon-dioxide build-up in Swildon's cave as a result of over-use of the available air by the constant passage of cavers! In general, the most dangerous practice of all is to rely on the air in small air-bells between sumps. Local cavers will be able to tell you if there is any bad-air problem in their area.

The saddest story about cave air comes from Mammoth Cave. In the last century a dozen people with lung problems decided to spend the winter in the cave for the sake of their health. They built stone huts in a capacious passage, and lit fires to keep themselves warm. The fumes killed them!

Burns

The most obvious source of underground burns is the acetylene flame of carbide lamps. Nasty if it catches your fingers, it can be disastrous if it burns your rope through.

Even nastier, because less easy to see, are leaks from rechargeable lamp-batteries. Lead-acid lamps can leak sulphuric acid; this attacks clothing and nylon ropes and harnesses. Nickel-iron and nickel-cadmium lamps can leak potassium hydroxide, which attacks polyester (terylene) ropes, and has a particularly virulent effect on flesh. The long-term answer to these chemical problems is the use of sealed-cell lamps which perform as well as existing miner's lamps. In the meantime, good maintenance and care is the best answer. Remember that lamp-batteries can take in water when submerged, and will tend to leak electrolyte afterwards.

Infections and accidents

Unfortunately, not every cave is wholesome! People have a tiresome tendency to regard caves as ideal dumping sites, and think that water travelling through the ground miraculously purifies itself. It doesn't. Unless you are certain that underground water is pure, do not drink it, and treat any cuts you receive as soon as possible. Cavers have contracted polio and leptospirosis (Weill's disease) underground. Histoplasmosis, a fungal infection of the lungs, can be contracted in bat-caves — usually, but not always, without long-term serious effects. Lastly, a strange condition named 'Mulu foot' has been reported, but only from Sarawak: the feet of cavers become pitted with small holes for a reason as yet unknown. Whatever 'Mulu foot' turns out to be, it is unlikely to appear anywhere other than in the tropics.

Dwelling unduly on the nasty accidents that *can* happen underground may get things out of perspective. With planning, due caution and appropriate experience, caving is reasonably safe. However, if things do go wrong you should be prepared, and ready to take the right action. Every incident is different, and you

Pictures from a real rescue. A boy who fell 45 metres down a disused shaft is brought to the surface by the Cave Rescue Organisation. He lived

will need to adapt the following check-list to the actual circumstances.

- Make sure the same thing does not happen to anyone else. Do not get clobbered by continuing stone-fall; fall down the same shaft; get swept away in the same streamway, etc., in your haste to help the victim.
- Get to the victim as safely as possible, using *all* necessary safety precautions.
- See that no further harm comes to the person who is hurt. If necessary, move the victim to a place where he is in no further danger; give first-aid, reassure him, make him comfortable and assess the situation.
- If necessary, send two of the best cavers in the party out for help. Self-help may be the best answer, and a rescue team should not be called out unnecessarily. If it is unavoidable, or you are not *certain* that you can get the casualty out under your own steam, send the two cavers out with a written message which states what has happened, where, to whom, when, and what help is needed. Include the name, address and age of the casualty. This may seem superfluous, but the police will insist on it. In the major caving regions there are special telephone numbers for cave rescue. However, if you are in any doubt, contact the police and ask them to contact the rescue team. In some parts of the world there are no cave rescue teams — bear this in mind when planning trips to remote parts!
- Continue looking after the injured person. Reassurance helps morale and the physical will to survive. Insulation prevents exposure. Where possible, clear low passages of stones, rig pitches, and so on to help the rescue team. Hand over to the rescue team when it arrives. Shocked and anxious members of the original team can be a liability. Unless rescuers are in very short supply, they are best kept out of the way.

Even practice rescue can be arduous — especially for the 'victim'

Damage to the cave

Caves can damage cavers, but much more frequently it is the other way round. And, whereas broken bones may take a few weeks to heal, a broken stalactite is broken for ever.

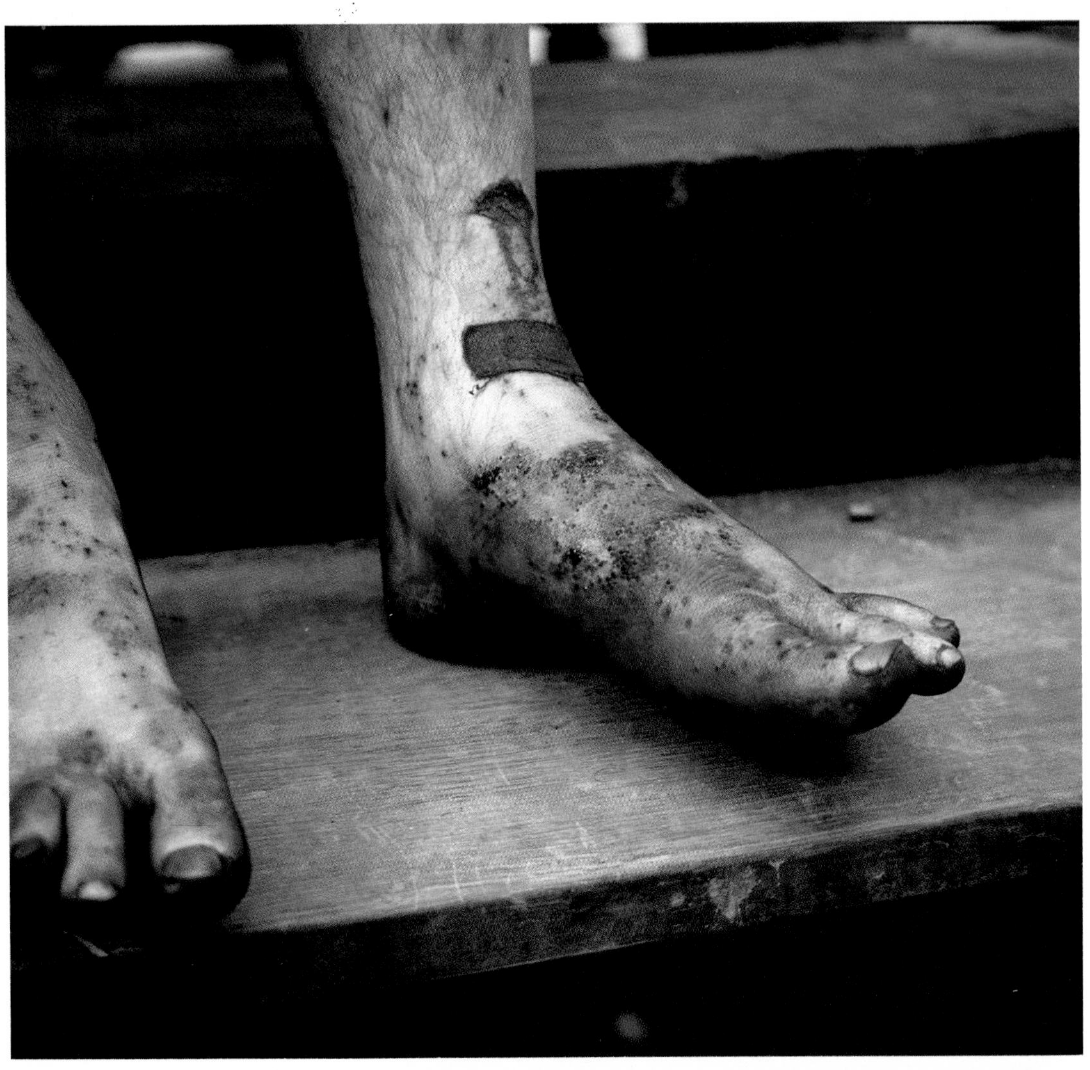

'Mulu foot' — mysterious hazard of the caves of Borneo

Cave centipede — an uncommon, but highly poisonous hazard! Mulu caves

Rubbish may be washed into a cave, or dropped in passing. Either way it stinks

When a cave is first discovered, it has a magical air. Fresh from its slow creation, it has delicate formations on roof and floor, and is a delight to behold. Aesthetics apart, it is a repository of knowledge about the past that speleologists are just beginning to learn how to interpret. Dating techniques based on the proportions of different isotopes of uranium found in stalactites (see page 36) can now take detailed chronology way back beyond the limits of tree-ring and radiocarbon dating. Layers of silt and mud on the floor of caves often contain a record of successive glaciations long since obliterated above ground.

The ecology of the cave, often made up of life-forms too small to be noticed by the passing caver, can be thrown out of balance by the disturbance of invading humanity. Every caver has a duty to safeguard the cave environment as carefully as he guards his own life, lest he destroys what he has come to see.

Never remove formations — do not even touch them unnecessarily. Try to follow only one track through wide passages. Formations are often taped off by cave discoverers, and such tapes should *always* be respected. Be especially careful of floor crystals and other ground deposits; they are particularly vulnerable. Do not think that the removal of just one stalactite from a display of hundreds will make no difference: if everyone thought the same way there would soon be none left. Footprints are unavoidable, but tread carefully and on the same track as others. Taking photographs gives you no excuse for dropping litter: there is no more disgusting sight underground than heaps of flashbulbs discarded among stalagmites.

Never leave debris of any kind. All too frequently users of carbide lamps leave piles of spent fuel lying about! This is a deplorable and totally unforgivable abuse of the cave environment. All waste, be it food scraps, spent carbide, or even a bit of ripped-off clothing, should be taken out, then taken home. Before condemning farmers and others for using surface holes as rubbish dumps, cavers must see to it that their own behaviour towards their caves is beyond reproach.

Take away your memories, and your rubbish. Leave nothing!

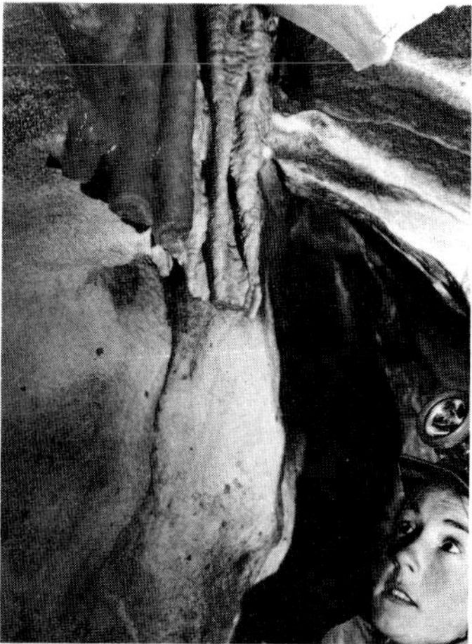

Once broken, stalactites will never be mended. Vandals snapped these formations off shortly after the Kingsdale Valley Entrance cave was found in 1967

Huntsman spider, Mulu, Sarawak

Finding and exploring caves

For your first year or two as a new caver you will probably be preoccupied with simply getting to the bottom of your cave and getting back out again. After this stage, many give up and try less arduous things, but cave-addicts usually continue to descend holes in the ground, more because they like being in and exploring caves than through any desire to pit themselves against them. Caving becomes a sort of three-dimensional physical and mental puzzle, with known sections of cave giving clues to the existence of unknown parts. There are a few places where exploration has been so intense that future discoveries are likely to be few — in Belgium and in southern England, for instance — but in most countries it can be confidently predicted that far more cave remains to be discovered than is already known, and even the 'played out' regions continue to throw up new surprises. The caver can still tread new ground. It is the dream of finding 'caverns measureless to man', and the very real prospect of finding something a bit more modest, that motivates the true caver.

It can be a matter of luck — the cowboy who in 1901 was riding past the undiscovered entrance of Carlsbad Caverns, New Mexico, when he had his hat blown off by wind belching from the ground is a classic example — but more often caves are found by understanding where to look.

Geological considerations

A study of geological maps will reveal tell-tale patches of limestone, usually shown in blue. Looking at a physical map of the limestone area may reveal evidence of streams suddenly disappearing and reappearing, or waterless closed depressions that in 'ordinary' terrain would hold a lake. Sometimes the evidence will not be so obvious — just a lack of surface drainage over a suspiciously large area. One thing is certain: the hardest part of a cave to find is usually its entrance!

In a 'new' area it is worth investigating all possible stream sinks and risings before combing the dry area in between: a systematic search of this type was done with enormous success by the Yorkshire Ramblers' Club around the Pennine hill of Ingleborough as early as the 1890s. Dyes such as fluorescein are used (with permission from the local water authority) to trace water flows from sink to rising. Such use gives an indication of the direction, the extent and the type of caves that the underground stream is traversing. Fast flow means open streamways; very slow flow probably means long sections of submerged passage. Only when

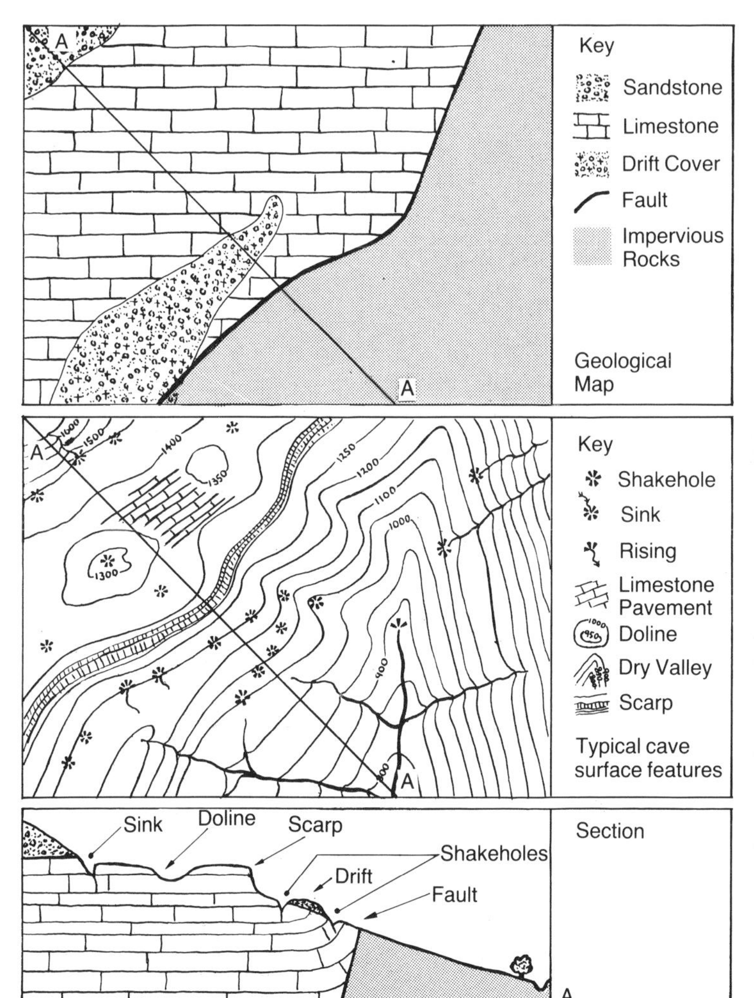

Typical cave-country surface features and underlying geology

Water spraying down shafts pulls air with it, creating natural ventilation

all stream entry and exit points have been exhausted is it worth moving on to the dry country in between. Even then, it is worth looking for fossil features of an earlier landscape — perhaps a downhill continuation of a 'dry' valley beyond the present sink, or an abandoned resurgence point now high above the valley floor. Look also for geological features such as faults or synclines.

Only when these leads have been exhausted is it worth combing entire mountainsides. In terrain with a covering of water-resistant material above the limestone — this covering can be soil, glacial drift, or thin sandstones or shales — the surface may be pitted with shakeholes. These look like bomb craters, and form where the surface water has been channelled through to dissolve the limestone below. The bigger the shakehole, the bigger the area it drains — but this also means that more mud and insoluble rocks have fallen to the bottom of the funnel. It is often the insignificant depression that yields the entrance you have been seeking!

In mountain 'lapiaz' — bare limestone slopes — the surface may be quite flat or contorted into all sorts of shapes. There may be no obvious entrances, or hundreds of shafts in every square

Sinks around Ingleborough. Caves are found where streams run off the upper shales onto the limestone.

Limestone pavement from above, showing drift cover with shakeholes in the upper section.

illustrated p.138
Cave in its undamaged state! Mendips, UK

kilometre. Aimless wandering may produce results, but it is better to choose an area, explore it carefully and methodically, and descend every shaft to locate the one in a hundred that actually 'goes'. A *small* (say 5cm high) painted reference at each entrance lets other cavers know a shaft has been looked at, and the results should be published so that others can refer to them.

In winter, cave-air rises to the top entrance of a cave; by

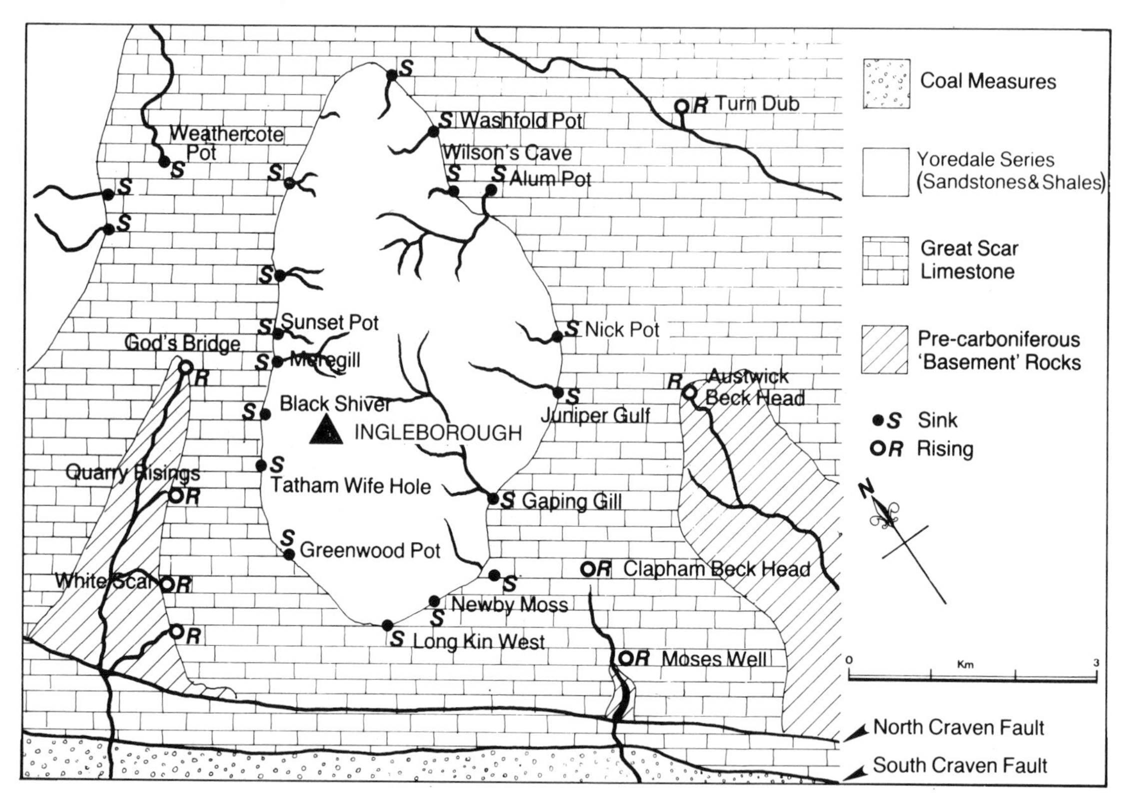

An actual sink and rising complex. Ingleborough, Northern Pennines, UK

Mountain lapiaz. Siebenhengste, Switzerland

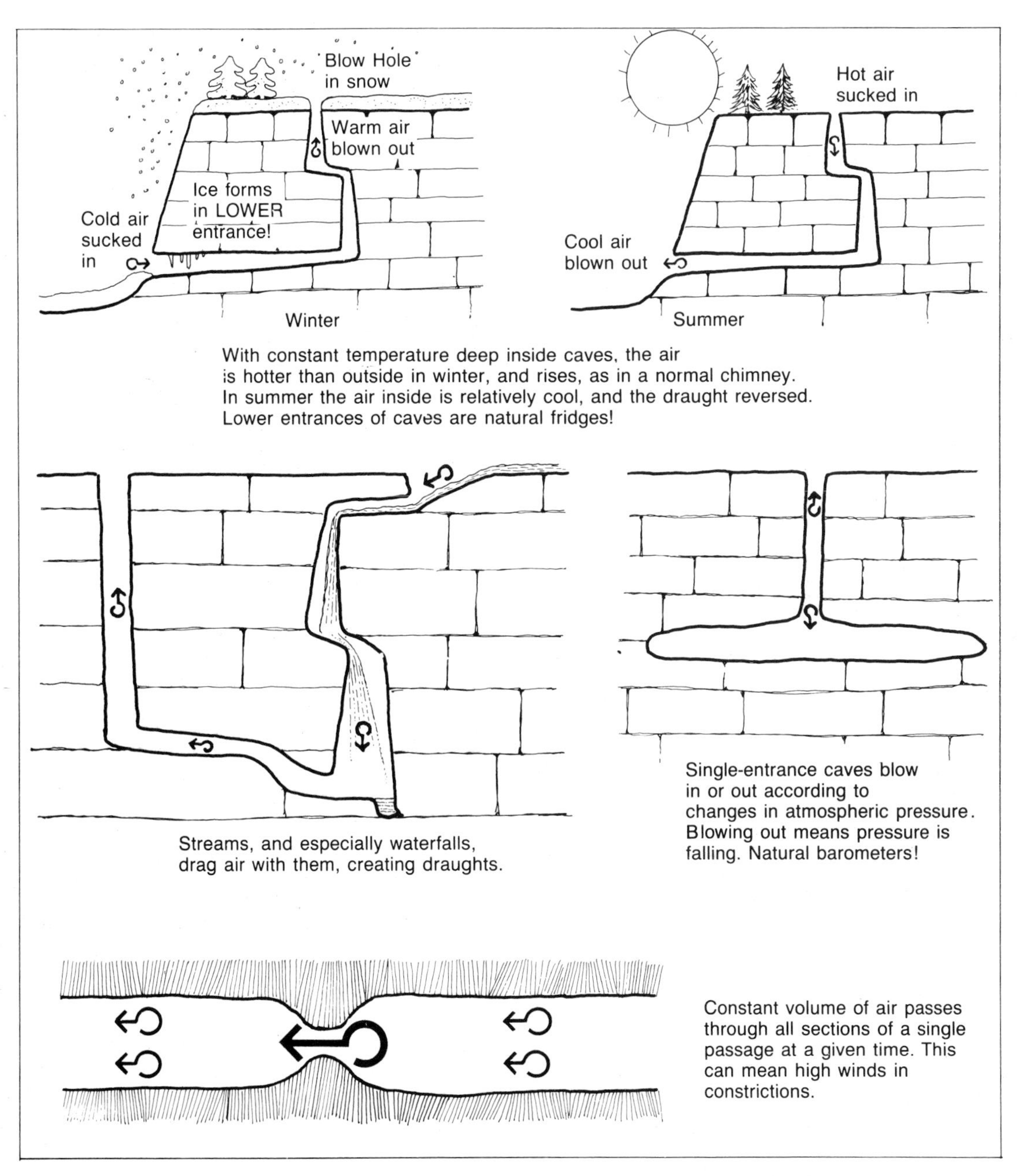

'Blow Hole'
in snow
Warm air
blown out
Ice forms
in LOWER
entrance!
Cold air
sucked
in
Winter
Hot air
sucked in
Cool air
blown out
Summer
With constant temperature deep inside caves, the air
is hotter than outside in winter, and rises, as in a normal chimney.
In summer the air inside is relatively cool, and the draught reversed.
Lower entrances of caves are natural fridges!
Streams, and especially waterfalls,
drag air with them, creating draughts.
Single-entrance caves blow
in or out according to
changes in atmospheric pressure.
Blowing out means pressure is
falling. Natural barometers!
Constant volume of air passes
through all sections of a single
passage at a given time. This
can mean high winds in
constrictions.

How caves breathe

searching a snow-covered surface you can occasionally spot a blow-hole melted by a rising draught, and so by-pass tedious hole-by-hole searching. Of course, this may be only the start of the story. A 'breather' in 10m of snow cover may be very difficult to mark so that you can find it for the actual exploration the following summer. Coloured tape tied to the nearest tree branch can be very elusive high amid the greenery! Conversely, snow can plug mountainside shafts so thoroughly that the only possibility of breaking through is to visit them just before the first winter snows, in late September or October.

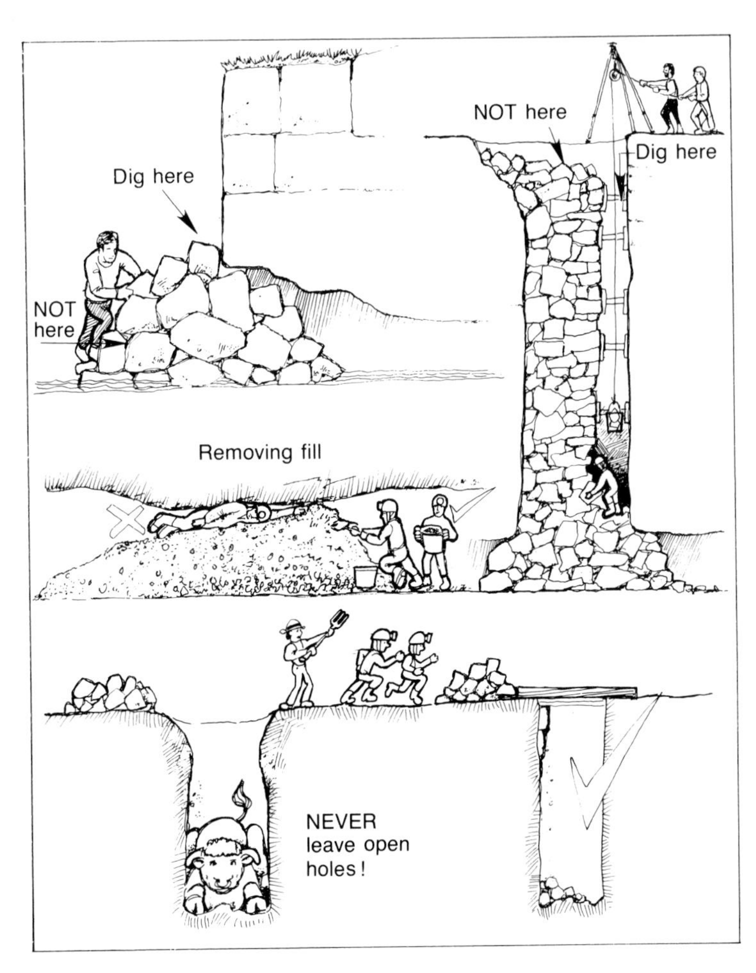

Digging for caves. Start digging at a solid face, preferably a solid roof or wall

Cave digging

If you are lucky, you may find a shaft or passage just waiting to be explored. More often you will have to dig your way in. Cave entrances have an infuriating habit of being collapsed, silted-up, or just too small to enter; and even once you are inside the cave further blockages are more than likely to interrupt progress.

The explorer must decide whether the blockage is worth digging. Is there a draught? Does the survey show a sudden end to a major passage that ought to continue? Has the passage actually become too small to follow, or is it simply blocked by redeposited calcite? Some cavers get a simple pleasure out of digging anything, but the rewards tend to come to the team that has chosen the best 'prospect'.

Cave digging is an art in itself, and one that some clubs concentrate on almost to the exclusion of caving proper! Just how you proceed depends upon circumstances, but there are some general principles. It is better to follow a solid rock surface, and preferably a roof, than to dig straight into a pile of loose boulders. If shoring is required, base it on a solid wall or floor. When clearing 'fill' out of a passage, it is best to remove enough to create a trench deep enough to kneel in: lying flat-out is not the easiest way to dig your way forwards. If the dig carries on for more than a

A cave dig in progress

illustrated p.142
Otter Hole, under Chepstow Racecourse, South Wales

Stages in the production of a cave survey

a. Grade 1 sketch survey made on initial trip without instruments

b. A page of notes from subsequent detailed surveying trip

c. Completed survey with plan, elevation, sections and key to signs used

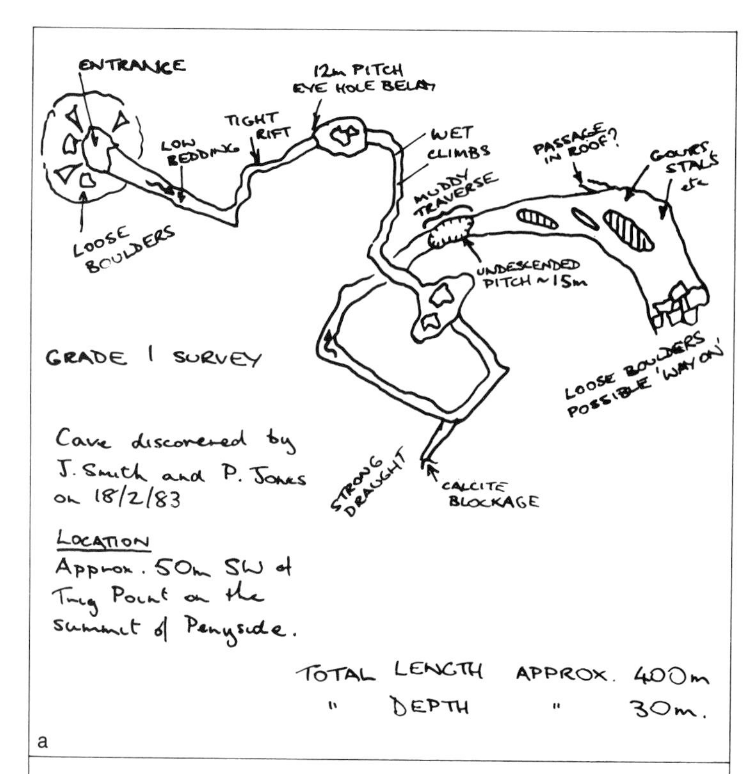

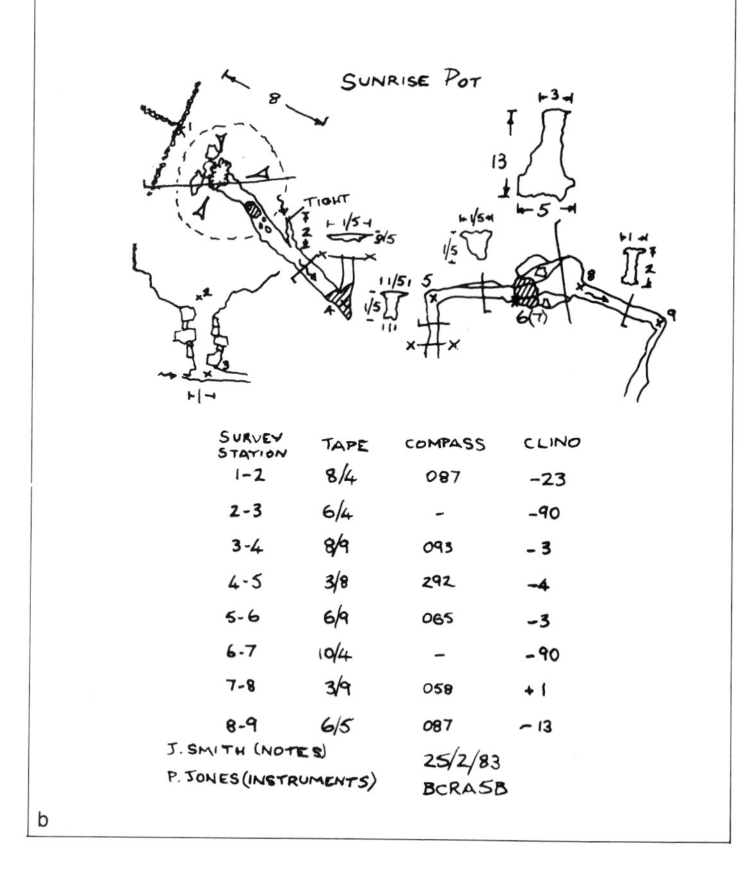

SURVEY STATION	TAPE	COMPASS	CLINO
1-2	8/4	087	-23
2-3	6/4	-	-90
3-4	8/9	093	-3
4-5	3/8	292	-4
5-6	6/9	065	-3
6-7	10/4	-	-90
7-8	3/9	058	+1
8-9	6/5	087	-13

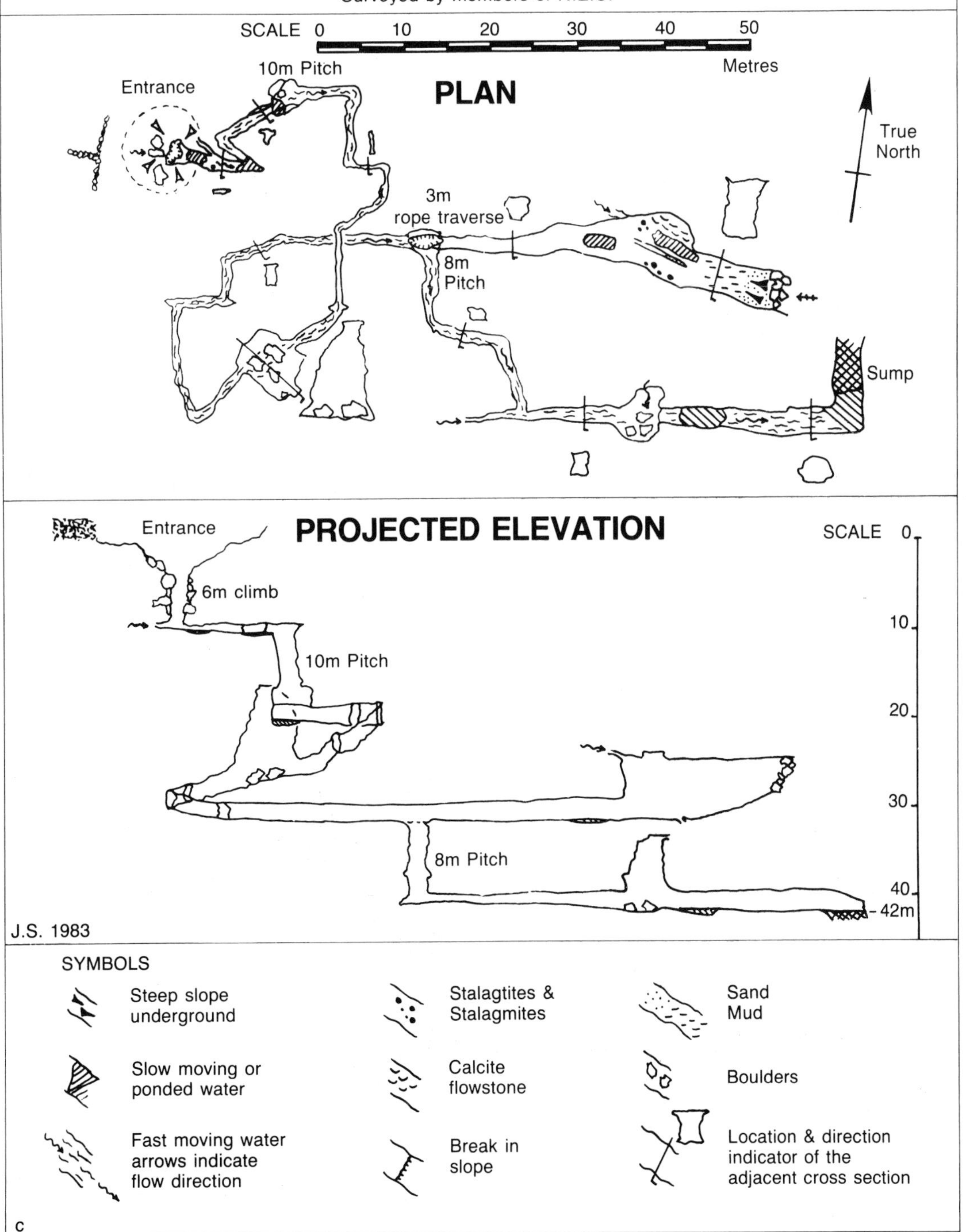

SUNRISE POT
PENYSIDE NORTH YORKSHIRE
National Grid Reference SJ 029416
SURVEY GRADE 5B
Surveyed by members of N.E.C.
SCALE 0 10 20 30 40 50
Metres
PLAN
Entrance
10m Pitch
True North
3m rope traverse
8m Pitch
Sump
PROJECTED ELEVATION
Entrance
6m climb
10m Pitch
8m Pitch
SCALE 0
10
20
30
40
-42m
J.S. 1983
SYMBOLS
Steep slope underground
Slow moving or ponded water
Fast moving water arrows indicate flow direction
Stalagtites & Stalagmites
Calcite flowstone
Break in slope
Sand
Mud
Boulders
Location & direction indicator of the adjacent cross section
c

Cave surveying. The distance, bearing and inclination of a series of 'legs' is the basis of the finished product — Sarawak Chamber

few metres, removal of the 'fill' becomes more of a problem than digging it loose in the first place. Simple buckets and manpower may have to be replaced by mining operations involving wheeled trucks on rails.

Explosives are widely used to remove short solid blockages and boulders too big to move. They are, of course, potentially dangerous, and should never be employed except by a trained person holding a police licence to possess explosives. Quantities of explosive should be kept as small as possible — certainly no more than a couple of hundred grams. The explosives should be placed in drilled shot-holes rather than plastered on the surface of the obstructing rock: this minimises the possibility of poisoning by fumes after the bang which, in non-draughting passages, or where the draught is blowing back towards the blasters, is a very real danger. In 1974 two very amateur cave blasters, trying to remove the roof of a sump in Cote Gill Pot in the Yorkshire Dales by exploding a large amount of home-made explosive, succeeded only in filling the entire cave with poisonous gases, killing themselves, and nearly killing some of the rescuers who went in to recover their bodies.

There is a code of ethics covering the digging and exploration of new caves:

- Always seek permission from the landowner or occupier.
- Cause minimum damage to the surrounding surface and the cave, and dispose of spoil tidily.
- Properly cover surface holes to prevent children or stock falling in.
- Never enlarge passages more than is essential.
- Never 'pirate' someone else's dig.
- Be particularly careful to get permission before prospecting in a foreign country.
- Leave your dig safe for others if possible.

The best way of achieving the last rule if you are leaving shoring in place is to employ stone-walling techniques, rather than use timber or metal poles, which rot.

Recording the 'new' cave

When the day comes that you break into a 'new' cave, you will have to make a record of it. Most important is a survey, or map, of the find.

A simple sketch drawn during the first trip, showing the main passages and junctions, will suffice for a time, but then an accurate survey needs to be made. Distances can be measured using a waterproof non-magnetic tape or a 'topofil' (a thread fed through a length-measuring meter and stretched between the survey points). Measurement of bearings requires an accurate sighting compass: Suunto or Silva models are the most widely used. The dip, or rise, of the survey line is measured with a clinometer — for accurate work the Suunto clinometer is best, but it is expensive; a protractor fitted with a plumb-bob may suffice.

Using these instruments, an accurate line-survey, showing length, direction and elevation, is made along all the passages. Detail is then added to the basic line, and the full survey will have

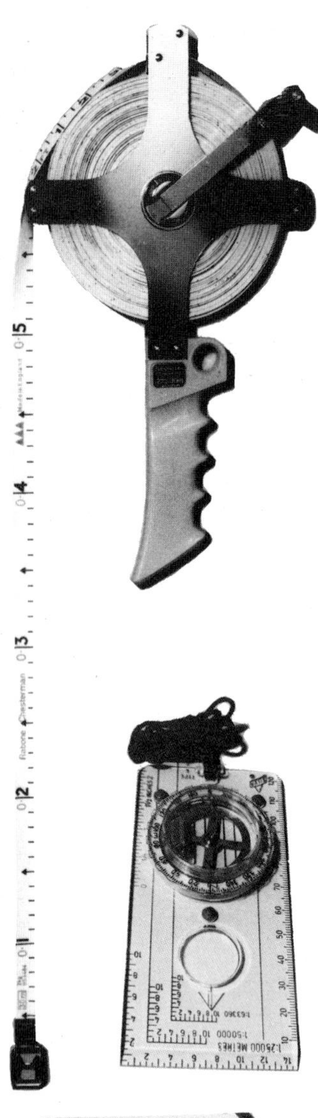

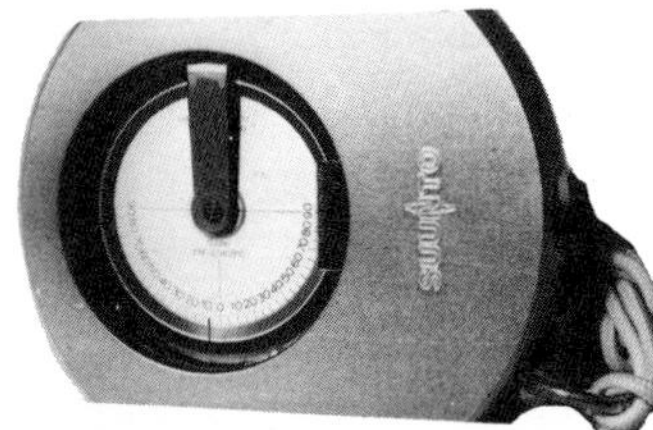

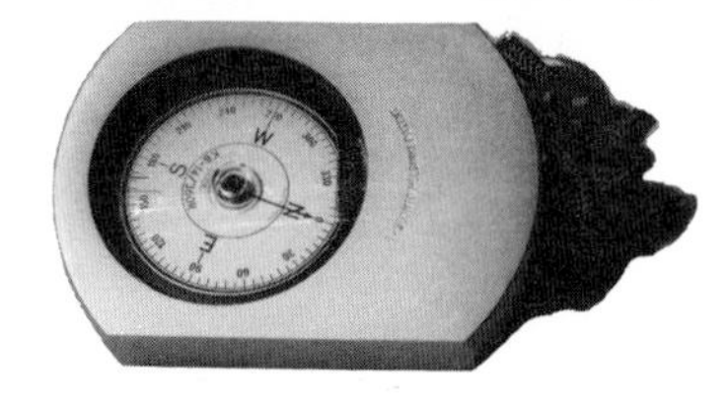

Cave surveying instruments: Sighting compass accurate to one degree, clinometer, waterproof measuring tape

a plan (looking down on the cave), an elevation (looking sideways), and cross-sections of typical passageways. There are conventional symbols for different features; those advocated by the British Cave Research Association are in wide international use. When your survey is completed it may reveal patterns of development, and possible links with other caves, that are not evident simply by looking at the cave.

You should write a full description of your 'new' cave to accompany the survey, and both should be published. If this can be done in a national caving magazine or journal, so much the better; records produced by the smaller caving groups tend to get lost in the mists of time.

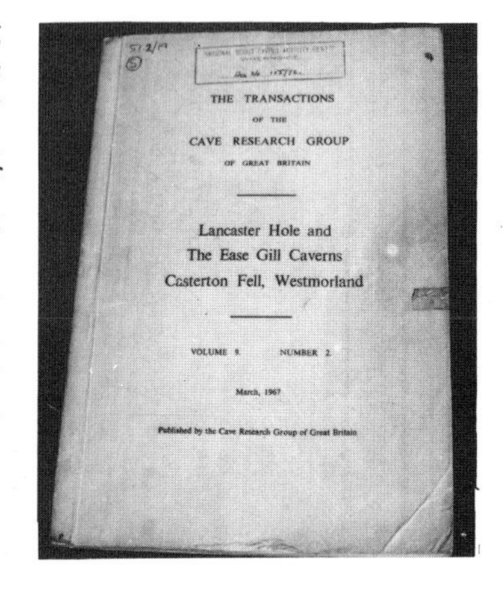
THE TRANSACTIONS
OF THE
CAVE RESEARCH GROUP
OF GREAT BRITAIN

Lancaster Hole and
The Ease Gill Caverns
Casterton Fell, Westmorland

VOLUME 9. NUMBER 2.

March, 1967

Published by the Cave Research Group of Great Britain

Cave photography

A description and map of the new find would be incomplete without a photographic record, so this is an appropriate place to discuss underground photography.

The first requirement is a suitable camera. Since there is no natural light underground, an automatic camera which adjusts to the amount of light available is largely irrelevant, and a manual-control camera is best. It must be robust, and should have an X-socket for flash-guns and a 'B'-setting to allow the shutter to be locked open with a cable release. Go for a 35mm camera rather than one of the modern mini-film models. There are some very compact 35mm cameras which produce good results, like the Rollei — alas, no longer produced — but most cave photographers tend to use a single-lens reflex model, the Olympus or Pentax being most popular with those who can afford them. For general underground shots, a fairly wide-angle lens is best — say 28–40mm focal length.

The second essential piece of gear is the flash-gun. Beginners may get the best results from bulb-guns. They are relatively powerful, and give a wider spread of light. However, good quality

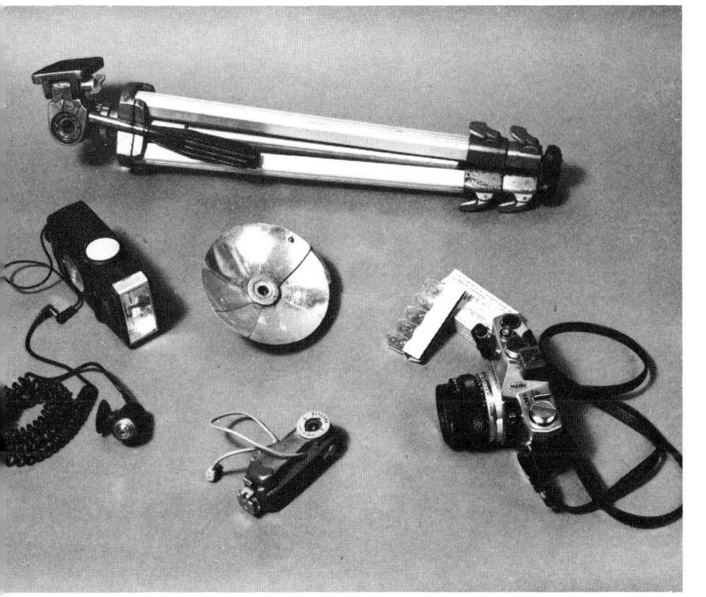

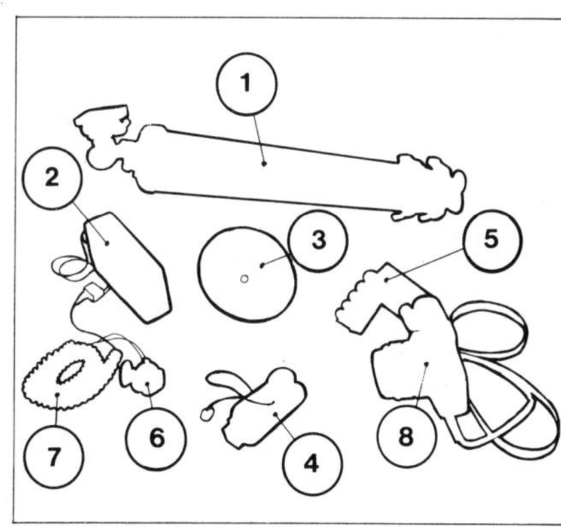

1 tripod
2 electronic flash-gun
3 bulb flash-gun (open)
4 bulb flash-gun (closed)
5 flash bulbs
6 slave unit
7 extension lead
8 35mm manual camera

For cave photography you need a robust 'manual' camera, flashguns (bulb or powerful electronic), extension leads or 'slave' units, and a tripod

Ogof Ffynnon Ddu streamway, front lit and back lit

illustrated p.150
Just imagine the difficulties the cave photographer has to take shots like this!

electronic flash-guns, provided they do not succumb to the damp, are more versatile, and are the usual choice of the more experienced. Although you *can* take pictures underground with an automatic camera fitted with integral flash, the results will not be good, and you will be limited to rather 'flat' close-ups — 'flat' because a flash-gun mounted on the camera does not produce shadows, essential if the shot is to have 'depth'; and close-up because the flash will not be powerful enough for anything else.

As an absolute minimum you need one flash-gun with a guide number of around 30 with 100 ASA film, which can be fired by an extension lead from the camera, or by pressing an open flash button on the flash-gun. For serious cave photography, at least two flash-guns should be carried, with extension leads for firing them and possibly 'slave' trigger units for firing all except one camera-linked flash. These cunning little devices plug into secondary flash-guns, and fire them when the first one is triggered from the camera. Using extension leads, and/or slave units, good results can be obtained from hand-held shots, but there is a limit to the amount of light that can be used.

A different sort of slave can be employed — in the form of other cavers. These have the advantage that they also carry your gear if spoken to nicely and promised the occasional shot with themselves in. However, you will also need to have a tripod, and only good-quality ones last long underground. With the camera securely mounted on a tripod, and the shutter held open with a cable release, you can collect an accumulation of image on the film as flash-guns are fired at will, provided that all cap-lamps are switched off while the shutter is open and that any models in-shot stand quite still.

How much light is required is not easy to judge. In the normal way, the guide number of the flash-gun, divided by its distance from the area to be lit, should give you the correct *f*-number to use on your camera. However, the rating of the flash-gun is for

1 ammo tin
2 cloth
3 padded lining
4 carrying strap
5 silica gel (desiccating agent)
6 tripod case

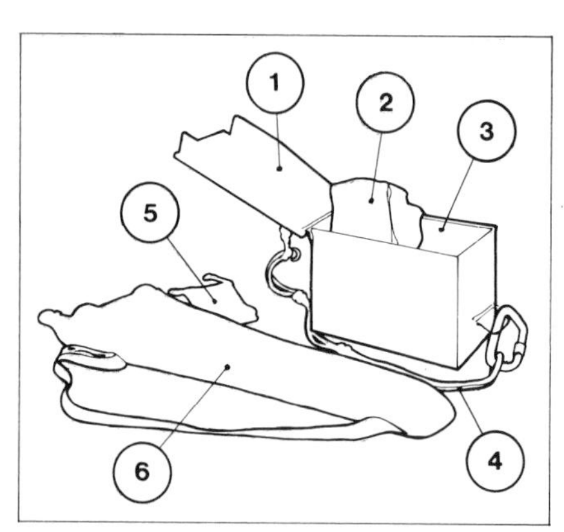

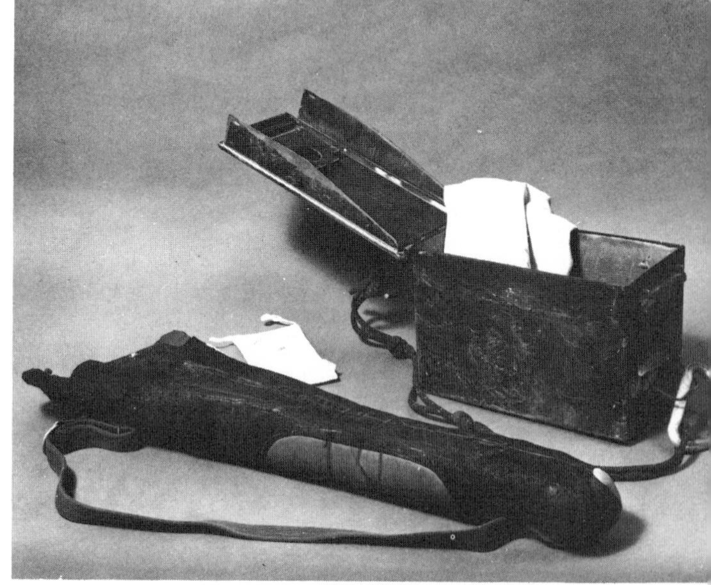

Gear protection: Silica gel for dryness, towel for padding and hand cleaning
Lined ammo tin and tripod cover for protection

averagely reflective surfaces. Most caves absorb much more light than ordinary surroundings, and so you must divide the given guide number by two, or, if the walls are very dark, three, to get reasonable results. This makes the use of reasonably 'fast' film almost essential for all but close-up shots. Film with an ASA of 400 will make most use of the light available. For big 'blow-ups' a good lens and 200 ASA film are preferable. (For black and white, Ilford HP5, Kodak TRI-X or Chromogenic film; for colour prints, Kodacolor; for colour slides, Ektacrome.)

Carrying your gear in and out of the cave needs just as much thought as the selection of the equipment itself. Mud, water, crashing and bashing combine to make formidable difficulties. These are best overcome by using an ex-army metal ammunition box. The inside should be lined with karrimat, or some similar material, for padding, and then with a towel. Camera, flash-guns, leads, film, etc., are then wrapped in the towel, which acts as extra padding — and also straightforwardly as a towel when you open the box with your wet hands. The boxes come in various sizes, and are available from specialist caving shops for a few pounds. A modest outfit will fit in a small (9×26×19cm) box; the photographer with everything may need a larger (14×28×19cm) box. In continually wet environments 100g or so of dried silica gel crystals should be carried in the box, wrapped in a stocking-bag. Heated in a medium oven, these crystals give out water; at cave temperature they absorb it, keeping their surroundings dry. When dry they are blue, when full of water they turn pink; this colour-change tells you when to revive them.

Tripods, of course, will not fit into ammo tins. Some cavers use a case made of heavy duty material, just to keep mud out, while others use a tube case made from plastic drain-pipe material. Jerry Wooldridge, who took many of the photos in this book, uses a case made from a discarded wet-suit arm!

Like most skills, cave photography cannot be taught from a book, but the preceding paragraphs should give you enough information for you to get hold of the right gear and use it underground. The rest is up to you!

Man, animals and caves

So far in this chapter I have referred to caves only in terms of naturally occurring geological features. However, they have often been used by animals and man for shelter and as living places, and can be extremely important sites for the archaeologist or palaeontologist. Cave sites have retained evidence from periods of which many or all surface remains have been removed — life in the ice ages, for example — and no such remains should be touched by cave discoverers. Many prime sites have been desecrated by mere collectors of artifacts.

This cave in Spain is still used by man — the even temperature is ideal for ripening cheeses!

Only in this century has the vital importance of the relative position and layering of deposits been fully understood. If you find a cave with evidence of animal or human remains, contact your nearest museum or university at once, and *do not touch them yourself*. To do so may be more than just a scientific heresy. In some parts of the world, cave entrances still contain sacred burial sites, and the local penalty for disturbing them may be death. This applies in the caves of Mulu, in Sarawak. Only weeks before I visited the area with a team of British speleologists, a burial site had been disturbed, and the old penalty was very nearly exacted. Needless to say, we went out of our way to avoid inviting the same fate!

The marine origins of limestone are clearly seen in this specimen from Cap Rhir, Morocco

illustrated p.155
Peak Cavern, Derbyshire, UK

Glossary of useful words and phrases

As in any other specialisation, caving has its own jargon. This tends to have an international flavour, with words in different languages for the same thing being interchangeable. Although puzzling to begin with, this does make it easier to converse with foreign cavers! Self-evident terms have been excluded.

Abseil. Descent of a fixed rope; *rappel* (French); rope-down.
Active. Description of a cave passage or cave with running water in it.
Anastomoses. Two-dimensional mazes of rambling and intersecting tubelets formed by solution along a plane of weakness in phreatic (*q.v.*) conditions.
Anchor. Sometimes used to mean belay (*q.v.*). More particularly, a threaded metal sleeve fixed in a drilled hole, and used as a belay in conjunction with bolt and hanger.
Aragonite. Relatively rare crystalline form of calcium carbonate, $CaCO_3$.
Ascend. In addition to the obvious meaning, this word is used to describe the process of climbing a fixed rope with the aid of jamming devices.
Aven. Shaft seen from below; sometimes restricted to those 'blind' at the top.
Bed. A layer of rock.
Bedding plane. The weakness separating two beds. It may be a fine crack, or several centimetres wide and filled with a different rock; in the case of limestone this is commonly shale, but it can be chert or coal.
Belay. A natural or artificial anchor point used to secure rope, ladder or caver. As a verb it means to use a belay. However, one does not belay to a belay, but simply belays.
Boulder-choke. Rock breakdown (*q.v.*) that completely fills a passage.
Bowline. Knot used to make a loop in the end of a rope.
Breakdown. Shattering of rock around passages and chambers along lines of weakness.
Breccia. Re-cemented breakdown. Often found in conjuction with faults.
Calcite. The common crystalline form of calcium carbonate, $CaCO_3$. Most caves formations are made of it.
Carbide. Short for calcium carbide, Ca_2C. Rock-like chemical which gives off acetylene when introduced to water. The fuel for carbide lamps.
Carbon dioxide. The gas which, dissolved in water, is responsible for the corrosion of limestone, and hence the formation of caves. Excess CO_2 can be a hazard in caves without an air-flow.
Carboniferous. Geological period which saw the laying-down of Britain's major limestones (around 270 million years ago), followed by shales, sand and grit-stones, and coal measures.
Cave. Any natural hole in the ground, though not usually applied to a vertical shaft. In France, a wine cellar! Also grotto, cavern, system, hole.
Cave cauliflower. Masses of white stuff that looks just like its name. A form of $CaCO_3$ 'fixed' from seepage water by bacterial action. Also called moonmilk (*Mondmilch*).
Chalk. Carbonate rock which has not re-crystallised, and retains its porosity.
Chamber. Part of a cave that is relatively large compared to the passages leading to and from it.
Chimney. Method of thrutching up a narrow rift.
Chockstone. Stone or boulder wedged between two walls.
Clinometer. Instrument which uses gravity to measure angles from the horizontal.
Clint. Surface limestone feature. A raised segment of rock bounded by corroded joints.
Corrasion. Erosion, or wearing away, by mechanical abrasion. Happens in caves when moving water bashes suspended or rolling sediments against walls and floors.
Corrosion. Erosion by solution.
Descender. Mechanical friction devices of many types for abseiling (*q.v.*).
Dip. The angle at which a bed of rock tilts from the horizontal.
Doline. Waterless closed depression. Dolines are major features of many karst (*q.v.*) landscapes.
Dolomite. Most limestone has a little magnesium carbonate in it. When the ratio of magnesium:calcium carbonates reaches about 30:70, dolomite results. Caves can form in it, although it is less soluble than limestone.
Dripstone. All types of calcite deposits resulting from precipitation from water.
Duck. Place where the cave roof dips to within a few centimetres of a water surface.
Eccentric. *See* Helictite.
Elevation. Sideways view; e.g., of cave on a survey.
Epi-phreatic. Part of cave that is periodically totally submerged.
Exposure. A subjective measure of the insecurity induced by a void beneath. Also, cooling of the body resulting from an inability of one's metabolism to generate heat as rapidly as the environment is removing it.
False floor. What is left when all the sediment in a passage is washed away except for its calcited top layer.
Fault. A line along which rock-slippage has occurred.
Fixed aid. Any mechanical aid to progression left in place in a cave.
Flood-pulse. Sudden peak stream flow that occurs after the onset of heavy rain or thaw.
Flowstone. Calcite deposited on the walls of passages.
Gours. Dams of calcite formed in streamways or on wet slopes. Running water containing excess CO_2 gives off more of this into the atmosphere when it is curving shallowly over a lip than when it is ponded. Hence more calcite deposition occurs on the lip, which builds up to become a dam. Like many other cave features, gours can be minute (less than 1cm) or enormous (5m plus).
Grikes. English name for the solution-widened joints bounding clints (*q.v.*).
Guano. Polite name for accumulations of bat or bird turds. Mined for fertiliser in some places. Best pepper is grown with the aid of guano.
Gypsum. Hydrated calcium sulphate, $CaSO_4.2H_2O$. A water-soluble rock which can yield extensive caves, notably in the Urals, USSR, and the Harz mountains, Germany.

Hading. Sloping — referring to a fault or rift.
Helictites. Cave formations growing at angles that defy gravity.
Joints. Cracks which are a ubiquitous feature of limestone, usually found at right-angles to the bedding.
Karren. German term, widely used, for all small-scale limestone surface features collectively. Similar to the French lapiaz (*q.v.*).
Karst. General name for limestone scenery and its typical landforms, after the Carso, a limestone area in the hinterland of Trieste, where they were first described.
Lagoon limestones. Limestone beds laid down in the 'quiet zone' between reef and shore.
Lapiaz. Bare, corroded limestone surfaces, usually in mountainous country.
Lifeline. Safety rope secured to the caver at risk, and belayed by another.
Marble. Limestone which has been metamorphosed by heat and pressure to obliterate its original structure.
Master cave. Stream passage to which all others lead.
Maypole. Sectional metal pole used with a ladder to gain access to an upper cave level.
Meander. Zig-zagging of a stream-cut passage due to greater erosion on the outer side of bends.
Metro. French term for large fossil conduit passage.
Moonmilk. *See* Cave cauliflower.
Ogof. Welsh for 'cave'.
Oxbow. Loop-passage abandoned by downcutting vadose streamway.
Pavement. General term for area of clints and grikes (*qq.v.*).
Pearls (cave variety). Round calcite stones which develop in pools under dripping water.
Permafrost. Condition in which a zone of rock below surface level is permanently frozen. In limestone, this seals the cracks that would normally drain water from the surface, so that during summer thaw conditions normal water-erosion processes mould the landscape.
Phreas. Permanently water-filled zone in a cave.
Phreatic. Term applied to a passage formed in water-filled conditions.
Pitch. Vertical shaft; more particularly, the route taken to descend it.
Plunge-pool. Deep hole scoured out at the base of a waterfall.
Poached egg. Initial stage of stalagmite formation — name is very apt!
Polje. Large closed valley which floods during the wet season.
Pothole or pot. Cave entered by a shaft; mainly vertical cave system; individual shaft. *See also* Rock-mills.
Productus. Large bivalve-shell fossil seen in carboniferous limestone.
Rappel. *See* Abseil.
Reef-limestones. Massive deposits with less regular laminar structure than lagoon limestones (*q.v.*). Often form separate hills (reef knolls) in the landscape.
Resurgence. *See* Rising.
Rift. Vertical or hading (*q.v.*) passage formed along a fault plane or a major joint.
Rig. Set of personal equipment for ascent and descent of ropes. *To rig:* equipping a pitch with ropes etc.
Rimstone pools. Pools surrounded by gours (*q.v.*).
Rising. Downstream end of a flowing sump. Also known as a resurgence.
Rock-mills. (French: *Moulins*) Rounded pits in stream beds eroded by the bashing of stones trapped in them. (Confusingly, called 'potholes' in conventional geography.)
Rope-down. *See* Abseil.
Run-off. The net effect of all the processes leading to drainage of a catchment by streams.
Scalloping. Current-markings in stream passage walls. The size of the indentations is inversely related to the speed of water flow.
Series. Part of a major cave system.
Shakehole. Crater left by collapse of soil/boulder clay/non-limestone cap-rock. The size is directly related to the thickness of the cover over the limestone.
Shale-band. Insoluble, friable layers found between beds of limestone.
Sink. Point of engulfment of surface water, an active shakehole. Also known as a swallet.
Siphon. *See* Sump.
Speleologist. Shortened to 'speleo': international word for caver or cave scientist. Equivalents: spéléologue (France), spelunker (US), etc.
Speleology. The study of caves.
Spreader. Short wire tether to fasten a caving ladder to a bolt belay.
Squeeze. Short constricted section of cave.
SRT. Single, or static, rope techniques.
Stalactite. Hanging calcite formation.
Stalagmite. Formation growing upwards under a roof drip or stalactite.
Straw. Hollow tube stalactite having the same diameter as a water drop.
Stylolite. Wavy line of impurities at an intersection of limestone beds. Indicates recrystallisation process.
Sump. Underwater passage, alternatively known as a siphon.
Swallet. *See* Sink.
Syncline. The base of a fold in rock strata. Cave formation is often channelled towards a syncline (e.g. Dan-yr-Ogof, in South Wales).
Tether. Wire rope strop used for belaying a caving ladder.
Traverse. A horizontal move at high level along a wall or rift.
Travertine. Limestone precipitated by evaporation, found mainly on the surface in hot, dry countries. May be associated with tufa (*q.v.*).
Troglobite. Animal which lives its entire life in the dark zone of a cave.
Tufa. Limestone precipitated from solution by plant action. It looks like the 'fur' in kettles.
Vadose. Refers to conditions obtaining when free-flowing streams with air-space above erode downwards.
Vauclusian. Refers to springs welling upwards steeply from deep phreas. The name comes from the Fontaine de Vaucluse, France, currently the world's deepest dived sump.
Water table. Upper limit of totally flooded zone underground. Well defined in porous rocks like chalk, it can be very variable and localised in limestone.

Resting, Arctomys Cave, Canada

Cave snake

illustrated p.159
Bridging the streamway. Lamprechtstofen, Austria

Further reading

The following list is a selection of works in print at the time of publication which the reader may find of interest and use. It is *not* comprehensive.

Caving Guides

Brook, D., Coe, R.G., Davies, G.M. and Long, M. H., *Northern Caves; Vol 1; Wharfedale and Nidderdale* (Dalesman).
Brook, D., Davies, G.M. and Long, M.H., *Northern Caves; Vol 2; Pen-y-Ghent and Malham* (Dalesman).
Brook, D., Davies, G.M. and Long, M.H., *Northern Caves; Vol 3; Ingleborough* (Dalesman).
Barrington, N. and Stanton, W., *Mendip: The Complete Caves and a View of the Hills* (Cheddar Valley Publications).
Irwin, D.J. and Knibbs, A.J., *Mendip Underground: A Caver's Guide* (Mendip Publishing).
Ford, T.D. and Gill, D.W., *Caves of Derbyshire* (Dalesman).
Stratford, Tim, *Caves of South Wales* (Cordee).

Exploration and explorers' stories

Eyre, Jim, *Cave Explorers* (Cordee).
Chevalier, Pierre, *Subterranean Climbers* (Mendip Publishing).
Casteret, Norbert, *Ten Years Under the Earth* (Mendip Publishing).
Lawrence, J. and Brucker, R.W., *Caves Beyond* (Mendip Publishing).
Farr, Martyn, *Darkness Beckons: History and Development of Cave Diving* (Diadem Books).

Technical Topics

Ellis, Bryan, *Surveying Caves* (British Cave Research Association).

Picture and large format books

Waltham, A.C., *World of Caves* (Orbis Publishing).
Deakin, P.R. and Gill, D.W., *British Caves and Potholes* (D.B. Barton).

Magazines

Caves and Caving, quarterly (British Cave Research Association). Available from caving shops or from B.M. Ellis, 30 Main Road, Westonzoyland, Bridgwater, Somerset.
Caving International, quarterly (Caving International). Available from caving shops or from Caving International, PO Bag 4014, Station C, Calgary, Alberta, Canada.
N.S.S. Bulletin, quarterly (National Speleological Society). Available from National Speleological Society, Cave Avenue, Huntsville, Alabama 35810, USA.
Spelunca, quarterly *(Fédération Française de Spéléologie)*. In French with an English summary. Available from F.F.S., 130 rue Saint-Maur, 75011 Paris, France.

Personal assistance

It would be possible to continue with 'useful information' ad nauseam. However this has not been done, and should the reader have any queries about any aspect of caving he or she is invited to contact the author at Whernside Cave and Fell Centre (National Cave Training Centre) Dent, Sedbergh, Cumbria, England, LA10 5RE.